AF419855

PRINCIPLES OF
MARINE VESSEL DESIGN
Concepts and Design Fundamentals of
Sea Going Vessels

PRINCIPLES OF
MARINE VESSEL DESIGN

Concepts and Design Fundamentals of Sea Going Vessels

Prasanta Kumar Sahoo

Florida Institute of Technology, USA

World Scientific

NEW JERSEY · LONDON · SINGAPORE · BEIJING · SHANGHAI · HONG KONG · TAIPEI · CHENNAI · TOKYO

Published by

World Scientific Publishing Co. Pte. Ltd.

5 Toh Tuck Link, Singapore 596224

USA office: 27 Warren Street, Suite 401-402, Hackensack, NJ 07601

UK office: 57 Shelton Street, Covent Garden, London WC2H 9HE

British Library Cataloguing-in-Publication Data
A catalogue record for this book is available from the British Library.

PRINCIPLES OF MARINE VESSEL DESIGN
Concepts and Design Fundamentals of Sea Going Vessels

ISBN 978-981-122-994-7 (hardcover)
ISBN 978-981-122-995-4 (ebook for institutions)
ISBN 978-981-122-996-1 (ebook for individuals)

For any available supplementary material, please visit
https://www.worldscientific.com/worldscibooks/10.1142/12090#t=suppl

Typeset by Stallion Press
Email: enquiries@stallionpress.com

Preface

THE DESIGN OF MARINE VESSESLS can be described as an amalgamation of marine design knowledge, which is passed to the next generation of Naval Architects and Marine Engineers. This was written with the intent to be the all-encompassing first step in marine vehicle design. Contained herein are chapters on nearly every ocean going vessel a young designer may encounter. The text presents clear definitions, examples and classification information.

Contents

List of Figures

List of Tables

List of Abbreviations and Symbols Used

Nomenclature

L : Length in general

LBP : Length between perpendiculars

DWL : Design Water Line

L_{WL} : Length Water Line

LOA : Length Over All

B : Breadth

B_T : Beam at Transom

T : Draft

D : Depth to weather deck

CL : Center Line of Vessel

C_B : Block Coefficient

A_w : Waterplane Area

C_P : Prismatic Coefficient

A_x : Section Area designed waterline for any section

C_M : Midship Coefficient

C_W : Waterplane area Coefficient

C_{VP} : Vertical Prismatic Coefficient

VBC/KB : Position of vertical center of buoyancy from keel

GM : Metacentric height

LCF : Longitudinal Center of Flotation

I_L : Longitudinal moment of inertia (about LCF)

I_T : Transverse moment of Inertia (about CL)

LCB : Longitudinal Centre of buoyancy

IMO : International Maritime Organization
  Δ : Displacement
  ∇ : Volume

List of Abbreviations

 FIT : Florida Institute of Technology
ITTC : International Towing Tank Conference
 ABS : American Bureau of Shipping
 LRS : Lloyd's Register of Shipping
DNV : Det Norske Veritas
  GL : Germanischer Lloyd
  BV : Bureau Veritas

Chapter 1

Design Process Overview

1.1. Introduction

The design engineering process is an intuitive and communal exercise.
By utilizing every available recourse, a product can be designed that
meets at the crossroads of cost-effective, operationally successful and
technologically advanced.

Every project conducted with a design process in mind should
start with a clear set of well-defined objectives from the client.
Without proper communication between the client, to management,
and design team, projects could go away wasting valuable time and
resources. A vessel, after all, is a collection of systems designed to
work in sync, with the potential of any one aspect disrupting the
entire system. With each change or adjustment, an evaluation must
be of connected systems made to confirm that there will be no
secondary effects throughout the vessel. The evolution of the vessel design is be a magical process as form and feature meet design
requirements transforming a design to a living ship.

1.2. Design spiral

The design spiral is an analogy for proper design considerations for
any engineering process with the pictured diagram expressly tailored
for Naval Architecture design. With each iteration of design, the
spiral starts again. Each sector is a representation of a critical design
area in a vessel, with each sector replaceable and adjustable with
each variation of design. As the design process progresses, the spiral
becomes tighter and tighter. This symbolizes the focus narrowing

1

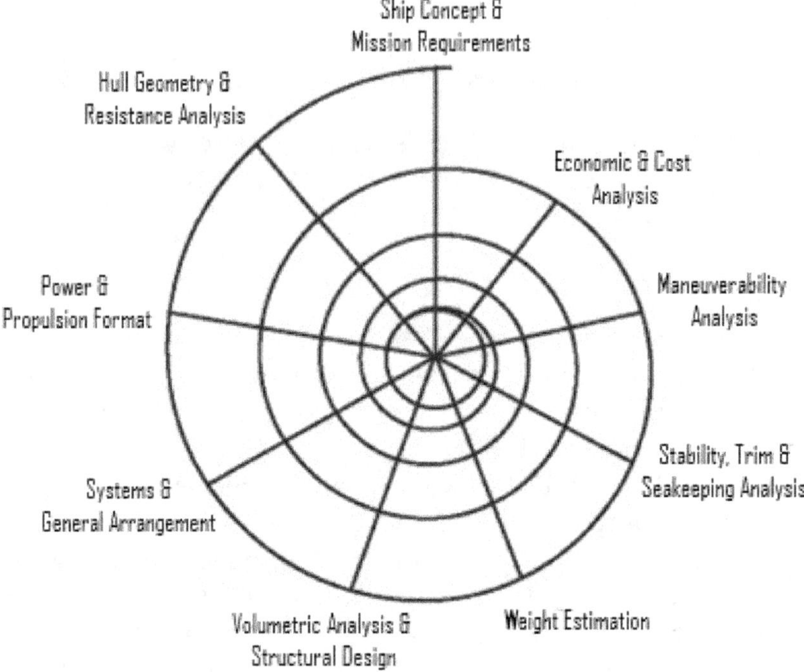

Figure 1.1: Design process spiral (design process and constraints)

down to the final design. More specific elements can be added to the design spiral to increase complexity or detail as well. In the case of a large cargo container vessel, the dimensional requirements that are centered on the length and draft determine the shipping routes a vessel can travel, therefore limiting the available work for said vessel.

1.3. Concept design

Once the primary design requirements are received, they must be translated into the systems in which a naval architect and engineer can work with. Establishing the feasibility and technological ability of the ship allows the mission requirements to be completed. By following the fundamentals of vessel building, each step should be complete;

hull type and form, primary dimension including length and draft, general arrangement, load displacement, payload volume requirements, form coefficient, required speed, and power specifications.

Primary design features are weight, VCG, and LCG estimates. These estimates are required for initial stability calibrations, however, without the inclining test, the stability calculations can not be finalized. In the inclining test, the quality of the design can be determined in that hull, structure, and outfitting. In addition to this, the test is also a precursor to the final sea trials preformed upon completion of the vessel.

1.4. Detailed design

Completion of the detailed design implies producing a set of working plans for fabrication, construction, and installation of all structural components required for use of the vessel. While this responsibility typically falls onto the shipyard or builder, the consultants or designers can also complete the plans. With the rise of internet connection, communication between the designers and fabricators has never been easier. Detailed plans and designs can be changed and uploaded over lunch much to the ire of the builders.

In the method tracking of design, the use of technical specifications can produce high quality control standards with the detailed drawing for every section. Considering technological advances in programs such as CAD software, computer-controlled milling and nesting of parts can greatly reduce material waste and cutting downtime. In addition, the CAD software allows complex piping and HVAC ducts to be designed as simple and hassle-free installation of the systems.

Much of the machinery and comments used in the vessel are commercially sourced from a variety of specialty manufactures. Fittings and power systems are designed and ordered months in advance to allow the manufacture lead time to align with build timeline. Therefore, in the preliminary design, it is vital to collect and amass all the lead times for equipment in order to maintain build scheduling. Any delay in the initial stages may throw the entire build time off and cause monetary penalties on the yard and designer.

In sum total, any vessel build project starts its life in an initial design phase. As this phase progresses, the design of the vessel is refined and approved. When more complex systems are added, consultants and communication with system manufacturers plays a significant role in the design. The liaisons responsibilities include reducing the strain on the initial designers and alleviating the wide range of specialties required to complete a vessel construction.

1.5. Mission requirements

What is the intention of this vessel under design, what is its mission and what is the intended performance? Do the potential future worth of the vessel justify the cost and operational expenditure. Before the first design spiral is started it is essential for the designer to consult the client in order to generate an expressly understood list of mission requirements for the vessel under design.

The mission outlined by the client can be very specific or very general and it is up to the designer to allow the client to fully relay their dream design. The primary controls of a vessels design and outfitting is economic ability. Through initial investment and future operational costs, running vessels is not a cheap task, forcing most naval design work to take place on commercial vessels. A good design should start with analysis of operational service. This is conducted by reviewing basic dimensions, propulsion and speed.

After consulting with the client and understanding what the primary intentions are, a mission statement or statement of requirements can be written. A concise statement presents the most professional report. However a degree of detail needs to be provided in order to give image to the vessel type and complexity of the system. In many cases, teams are used to develop these initial requirements and the *mission statements* in these cases may function as a brief defining the goals of the project. The mission statement provides a tricky situation where it the brief should not inhibit the design creativity, however the statement should provide clear guidelines to prevent potential questionable design decisions.

When the mission statement and design spiral are used in concert, the design process can function smoothly and efficiently. With each iteration and review of the design spiral against the overarching mission goals, no discrepancies will arise from the requirements.

1.6. Mission statement example: concise

A concise example of a mission statement for an Oceanographic Research Vessel of 75 meter class is given below. These types of vessels are typically complex and the missions it is deployed for may change and be adjusted on the fly. The example provided below from the initial design brief was put together using the consultancy and design requirements put out from the client.

1.7. Mission statement example: detailed

A detailed mission example is best applicable when the customer is sure of the requirements. The detailed statement should include; minimum service speed, range, maneuverability, intended routes, vessel class, port of registry and manning. When these requirements are set in stone a detailed version can be released. However it is vital that the customer understands that these requirements should not be changed or adjusted without serious consideration.

Mission Statement

The Vessel is to provide a dedicated platform for ocean exploration to a working depth of 10,000 meters. The vessel will explore biological, geological, physical aspects of oceanographic research, and the surface-support of manned and unmanned underwater vehicles, at a maximum practical range from the vessel's base port in Sydney, Australia. Flexibility in operations is to be permitted by the modular design of the systems as required for individual and varied voyage profiles and missions. The vessel is to accommodate a scientific complement and crew necessary to the needs of research both present and future with appropriate endurance to allow global deployment possibilities and potential to be utilized continuously.

Figure 1.2:   Mission statement, concise (IMO.2014)

100,000 tonne dwt 'GENCO' Bulk Carrier

The project is to complete the development of a design for a bulk carrier with a minimum cargo deadweight capacity of 100,000 tonnes, intended for the transport of alumina from Gove in Northern Territory, Queensland to Portland, Victoria. To minimize transits in ballast, the vessel is to possess the additional ability to convey bulk gypsum, coal sinter and other fine-grain bulk cargoes from Victoria and NSW ports to both Brisbane and Gladstone. Draft constraints are to be determined by the vessel's required ability to negotiate the inner channel of the Barrier Reef between Gladstone and the Gannet Passage (west of Thursday Island). Overall length is to be governed by the upper limits for berthing facilities at the ALCOA Weipa terminal.

Specific Requirements

Minimum service speed:	*16 knots in condition up to sea state 5 inclusive.*
Minimum range:	*4,500 nautical miles*
Maneuverability:	*Autonomous berthing/deberthing capacity under conditions of Beaufort Force 6 beam winds at designed load and ballast drafts. Full-service speed maneuvering capacity within the Great Barrier boundaries, and positioning/heading maintainability in cyclonic conditions.*
Refueling Ports:	*Brisbane, Newcastle, Melbourne*
Class:	*Lloyd's Register (100 A1) LMC, UMS, Bulk Carrier*
Port of Registry:	*Victoria, Australia*
Manning:	*Compliance with AMSA and IMO recommendation (minimum complement 1)*

Figure 1.3:   Mission statement, detailed (IMO.2014)

1.8. Design fundamental considerations

The mission requirement fundamentals are prescribed directly from the design considerations from the client. In most cases following the design spiral, the first section of the appropriate hull form for the designed specifications. The following requirements for power, propulsion and maneuverability then leads into engine specifications. The materials and port of call are the last sections before the design spiral repeats and narrows the specifications. As the specifications narrow, avoid becoming too immersed in the specifics as this can lead to issues with meshing the full system. In practice a power system should not be selected before the requirements for vessel speed and reliability are known. In addition, designing the intricate hull form that will not be able to accommodate the size of the cargo will generate tremendous waste in design time and efforts.

1.9. Specific design requirements

Using the examples provided above, specifically the cargo ship. A detailed mission statement might provide general requirements and performance criteria. Manning, classification, maneuverability characteristics, range and cruising speed. Additional requirements are stated as well including, but not limited to, fuel consumption and refueling ports. The fuel types are a much different matter, primarily designed by the naval architects. The decision is made by comparison of machinery systems using residual fuel oil in conjunction with expensive filtration devices, or marine diesel oil with low maintenance systems and simple filtrations devices.

1.10. Operational upkeep

During detailed design, it is beneficial to envision the work profile of the vessel. Depending on type, the operational timeline of a vessel can differ by a round trip-voyage, a calendar year or even once a day trip. With these trips, speed and power need to be considered, forming the following list of aspects:

- Hull Lines (classic, specific, exotic)
- Propulsion (class, style, speed)
- Fuel Considerations (type, availability, ports of call)
- Engine (number, sizing, type)
- Power systems (generator, sizing, type)
- Storage (provisions, range, supplies)
- System Load (electrical, hydraulic, communication)

Additional considerations should be focused on as well. Sea conditions, endurance, consumable reserves, and specified power should all be centered around adverse condition. For ocean going vessels this requirement for reverses is often quoted at three days on most vessels with the ability for it to be raised depending upon the duration of the voyage.

1.11. Manning specifications

A good crew is vital to the ability of the vessel to preform its operations underway. The governing bodies of maritime operations have seen fit to institute a number of requirements for the minimum and qualification standards for crews dependent on vessel class and flag state. In practice naval architects should try to reduce crew requirements and ease the manning of the vessel. As necessary, a designer should have a practical and detailed understanding of operations conducted aboard, and what each member of the crew might be expected to accomplish for the entirety of vessel operations.

Manning level has several implications on design including, accommodation, general arrangement and other operation features expected. More about each specific type of reduced manning can be found in the chapters dedicated to vessel type.

1.12. Classification societies

Society	Abbr.	Origin
International Maritime Organization	IMO	United Nations
American Bureau of Shipping	ABS	United States of America
Lloyd's Register of Shipping	LR	Britain
Germanischer Lloyd	GL	Germany
Korean Register of Shipping	KR	Korea
China Classification Society	CCS	Peoples Republic of China
Nippon Kaiji Kyokai	NKK	Japan
Det norske Veritas	DnV	Norway
Registro Italiano Navale	RINA	Italy
Bureau Veritas	BV	France

The societies have committees of every major faucet of marine engineering and naval architecture. More information can be found from each group site.

Chapter 2

General Cargo Ship

2.1. Introduction

Any ship or vessel that transports heavy goods and materials from one port to another is called a cargo ship. A general cargo ship is extremely adaptable and can be used to transport virtually every form of dry non-bulk cargo, from railway lines to agricultural machinery. A distinct feature of general cargo ships is that they normally have their own gear. This loading and unloading equipment allow versatile ships to trade with smaller ports and terminals that lack in shore side facilities. While these ships carry nearly every type of cargo, they are often loaded with abnormal cargo other vessels could not accommodate. During lean times general cargo ships can easily turn their hand to carrying containers, bulk or bagged cargo.

Modern dry cargo ship designs maximize hold space. Older designs typically have three holds forward of the superstructure and two aft. A typical mid-size modern ship may have five or six holds; three or four forward of the machinery space and superstructure, and one or two aft. The machinery spaces and superstructure are usually located about three quarters aft. Holds aft of the accommodation and machinery spaces improve the trim of the vessel when partially loaded, and provide the ship with sufficient draft aft for stability and propeller immersion. Small freighters, 150–180 meters in length, often have machinery and accommodation spaces aft of all cargo holds. Research has shown, deadweight of modern general cargo liners ranges from 9,000 to 25,000 tons; design speeds range from 17 to 22 knots.

Figure 2.1:　Genco spirit (Genco Shipping.2015)

Table 2.1:　Design specifications for bulk carriers [Maritime Studies South Africa]

Ship Type	Length (m)	Deadweight
Capesize Bulkers	300–400	150,000–400,000
Panamax Bulkers	200–240	60,000–80,000
Handysize Bulkers	150–180	25,000–40,000
Handymax Bulkers	160–180	40,000–60,000

2.2.　Specifications of the design

From the studies of experimental data, the speed-to-length ratio is generally 0.87 or less as higher ratios are usually not economical. Laden drafts are as deep as channels to the intended terminal ports allow, typically in the 7.92 m to 8.84 m range. Hull depth is selected to provide the desired draft and satisfy statutory freeboard requirements. Depth of the double bottom is kept low to maximize cargo space. One or more tween decks may be fitted to facilitate flexibility in cargo loading and unloading, cargo segregation, and to improve stability. There may be watertight doors in the bulkheads on the tween decks levels. Denser cargoes are carried in the lower holds with high stowage factor products normally stowed in the tween decks. Stowage factor refers to the ratio of cargo weight to stowage space required under normal conditions, typically referred to in m^3/t. Refrigerated spaces may be built into the tween decks. Ship designs for a specific trade strive for "full and down" operation; the ship's freeboard is down to her load line with cargo cubic fully occupied.

For a given trade, hold spaces are usually designed so that the ratio of bale cubic to deadweight is 10 to 15 percent greater than the overall stowage factor of the goods carried to allow for more rapid cargo handling and broken stowage — the spaces between and around cargo units, including dunnage, and spaces not available for cargo stowage because of physical obstructions or ventilation and access requirements.

Holds are sized and provided with cargo gear to limit the amount of cargo cubic per stevedore gang to about 1699.0108 cubic m. Holds in the midbody are therefore usually shorter than those further from midships. The conflict between the desire to shorten holds and the length required by cargo gear and hatches sometimes dictates the assignment of midships spaces, when cargo is unavailable the space is predominantly assigned to machinery or to fuel, cargo, or ballast deep tanks rather than holds.

Hatches are designed as large as possible without compromising hull strength (the main or second deck is normally the strength deck)

Figure 2.2: Ships gear (Great Seafarer Training.2015)

to reduce the requirement for horizontal movement of cargo within the holds. Hatches served by two sets of cargo gear generally measure 6.096 by 9.144 m or larger. Hatches on older ships are generally smaller than those on newer ships, reflecting dew design philosophy.

Hatches are surrounded by coamings to reduce the risk of flooding in heavy seas. Covers are usually constructed of steel (or wood on older vessels). The main deck plating between hatches is not effective in providing longitudinal strength, and is sized to carry fairly light local loads. The deck plating outboard the hatches is therefore much heavier, often exceeding five-eighths inch in thickness.

Cargo gear is designed for speed and flexibility for handling break bulk, palletized, or container cargo. Various combinations of derricks, winches, and deck cranes are used for the handling of cargo. Cranes are fitted on many vessels to reduce manpower requirements. Some ships have special heavy-lift derricks that may serve one or more holds. Booms are rigged for either yard and stay (burton) or swinging-boom operation.

Various combinations of derricks, winches and deck cranes are used for the handling of cargo. The economic factor is of prime importance in designing a merchant ship. An owner requires a ship which will give him the best possible returns for his initial investment and running costs. This means that the final design should take into account not only present economic considerations, but also those likely to develop within the service life of the ship.

With the aid of computers it is possible to study a large number of varying design parameters and to arrive at a ship design which is not only technically feasible but, more importantly, is the most economically efficient.

The Figure 2.3 shows the midship section of a General Cargo Ship. These sections are designed explicitly in the standards of the governing shipping bodies (IMO, ABS, LR, GL) to name a few. These restrictions are imposed to maintain a standard for shipbuilding. The international conventions allow any builder and ship operator to trust the vessel under steam. The member states are committed to upholding the standards for design, vessel strength, material selection and handling of cargo. By forming this standardization the shipping industry has exploded in growth with no stoppage in sight.

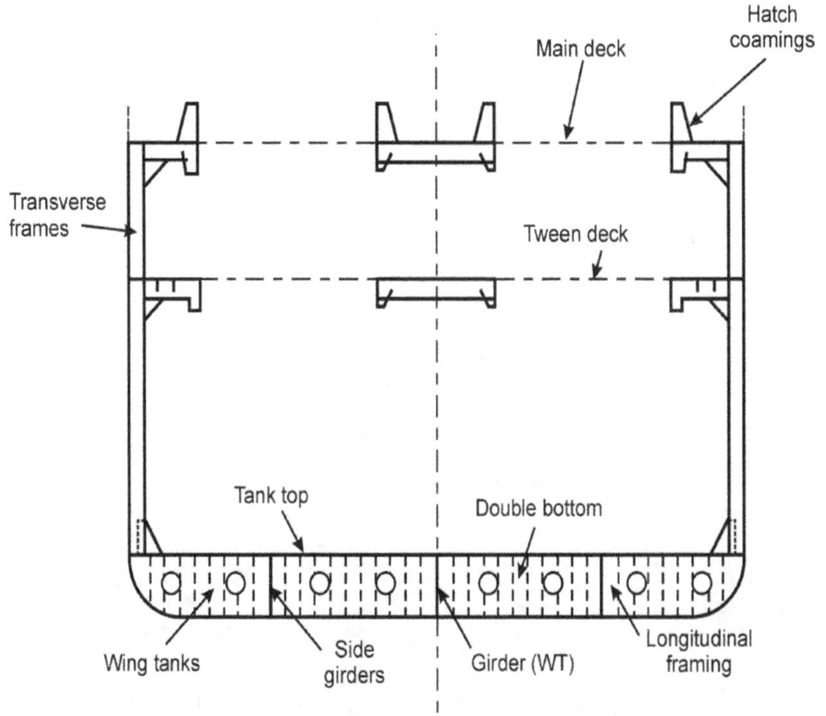

Figure 2.3: Midship section of a general cargo ship [Wikimedia Commons (2011)]

2.3. Important parameters

The basic factors in determining the main sizes are summarized in the following:

- Length L: This is a function of displacement and speed. It has a significant influence on the weight of steel structure and accommodation/outfitting, hence on the construction cost. In addition, it has great affect on both the ship's calm water resistance and seakeeping performance (motions, accelerations, dynamic loads, added resistance, and speed loss in seaways).
- Block coefficient C_B: This is a function of the Froude number and is influenced by the same factors as for the length L.
- Displacement Δ: A preliminary estimate of displacement can be made from statistical data analysis, as a function of deadweight capacity.

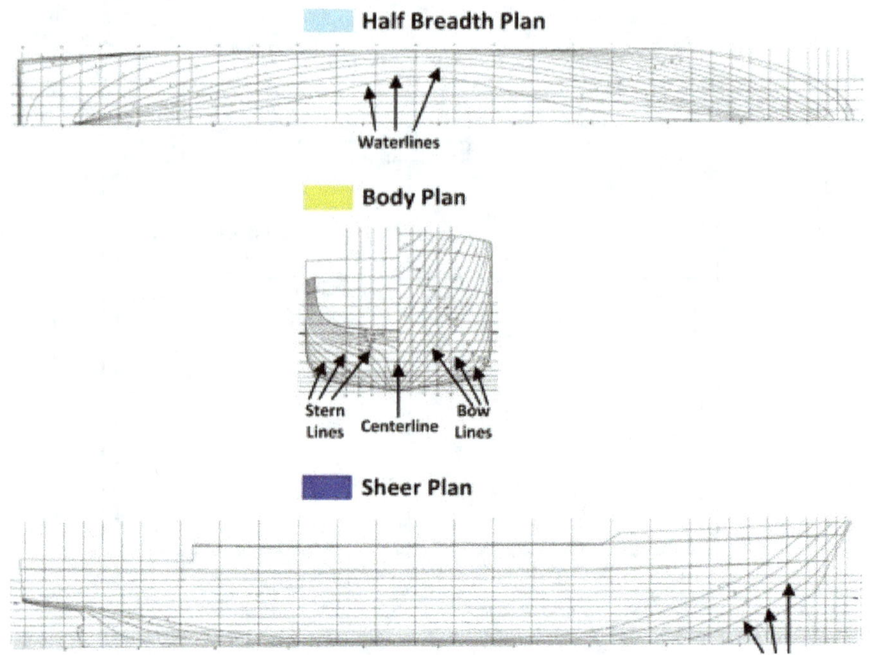

Figure 2.4:   Lines plan bulk cargo vessel (Officer of the Watch)

- Beam B, Draft T, side depth D: The determination of these dimensions is actually coupled and is affected by the following basic factors:
  - hold volume (D)
  - stability (B)
  - required freeboard (D, T)
  - safety against flooding and capsize (B, D, T)
  - propulsive and maneuvering devices (T),

The main dimensions L, B, and T are often affected as well by the topological limits of the route, that is, the dimensions of canals, ports, channels, and confined waters that the under-design ship needs to pass through. Mostly the restrictions are referring to allowable drafts. Table 2.2 shows some of the dimensions.

Some of the parameters of a Cargo Ship are given below in Table 2.3.

Table 2.2: Typical dimensions of well-known canals and channels (maximum allowable ship dimensions) [A. Papanikolaou, Ship Design, Springer]

Water Body	Length (L)(m)	Breadth (B)(m)	Draft (T)(m)
Panama Canal	<289.56(general merchant ships) <299.13 (passenger and container ships)	<32.31	<12.04 (for tropical freshwater)
Suez Canal	No limit on length	<71.02	<10.67 (concerning stern draft in ballast condition) <12.80 (max allowable T for B<47.55 concerning full load voyages southbound) <16.15 (max allowable T for B<42.67 concerning full load voyages northbound)
Canal St Lorenz (North America-Canada Great Lakes)	<222	<23	<7.6
Northeast Sea Channel (Nord-Ostseekanal-Northern Europe)	<315	<40	<9.5
Malacca Straits	—	—	<25

Table 2.3: Hydrostatics of a cargo ship [Dr Prasanta Sahoo, "Preliminary Ship Design", Internet data]

Parameter	Value
C_B	0.7
C_P	0.61–0.76
C_M	0.97–0.98
Velocity Range	13.5–15 knots
B/T	$2.25<=B/T<=3.75$ (conventional monohull)
	~5 (heavily draft limited vessels)
T/D	0.7 for type B freeboard
	0.7–0.8 for B-60 freeboard

The expression for draft is give as $T = 4.536[\text{dwt}/1000]^{0.290}$ m.

2.4. Initial estimate of stability

2.4.1. *Vertical centre of buoyancy KB*

For a general cargo ship C_M is >0.9, hence according to Posdumine and Lackenby (year unknown),

$$\text{KB}/\text{T} = (1 + C_{VP})^{-1}.$$

Regression formulations are as follows:

$$\text{KB}/\text{T} = 0.90 - 0.36\,C_M$$

$$\text{KB}/\text{T} = (0.90 - 0.30\,C_M - 0.10\,C_B),$$

$$\text{KB}/\text{T} = 0.78 - 0.285\,C_{VP}$$

where KB: Vertical centre of buoyancy and C_{VP}: Vertical Prismatic Coefficient.

2.4.2. *Metacentric radius: BM_T and BM_L*

Moment of Inertia coefficient C_I and C_{IL} are defined as

$$C_I = I_T/\text{LB}^3$$

$$C_{TL} = I_L/\text{L}^3\text{B}.$$

Table 2.4: Equations for estimating water plane inertia coefficients [Dr Prasanta Sahoo, "Preliminary Ship Design"]

Equations	Applicability/Source
$C_1 = 0.1216\,C_{WP} - 0.0410$	D' Arcangelo transverse
$C_{IL} = 0.350\,C_{WP2} - 0.405\,C_{WP} + 0.146$	D' Arcangelo longitudinal
$C_I = 0.0727\,C_{WP2} + 0.0106\,C_{WP} - 0.003$	Eames, small transom stern (2)
$C_1 = 0.04\,(3C_{WP} - 1)$	Murray, for trapezium reduced 4% (17)
$C_I = (0.096 + 0.89\,C_{WP2})\,/\,12$	Normand (17)
$C_I = (0.0372\,(2\,C_{WP} + 1)^3)\,/\,12$	Bauer (17)
$C_I = 1.04\,C_{WP2})\,/\,12$	McCloghrie + 4% (17)
$C_I = (0.13\,C_{WP} + 0.87\,C_{WP2})\,/\,12$	Dudszus and Danckwardt (17)

The formula for initial estimation of C_I and C_{IL} are given below in the Table 2.4.

2.4.3. *Transverse stability*

$KG/D = 0.63$ to 0.70 for normal cargo ships

$$KM_T = KB + BM_T$$

$$GM_T = KM_T - KG$$

Correction for free surface must be applied over this. Then,

$$GM'_T = GM_T - 0.03\,KG \text{ (assumed)}.$$

This GM'$_T$ should satisfy IMO requirements.

2.4.4. *Longitudinal stability*

$GM_L \sim BM_L = I_L/\nabla = (C_{IL}.L^3.B)/L.B.T.C_B = C_{IL}.L^2/T.C_B$

$MCT\,1\,cm = \nabla \cdot GM/100 \cdot L_{BP} = L \cdot B \cdot T \cdot C_B \cdot C_{IL} \cdot L^2/100 \cdot T \cdot C_B.$

$L = C_{IL} \cdot L^2 \cdot B/100$

2.4.5. *Longitudinal centre of buoyancy*

The longitudinal centre of buoyancy LCB affects the resistance and trim of the vessel. Initial estimates are needed as input to some resistance estimating algorithms. Likewise, initial checks of vessel trim require a sound LCB estimate. In general, LCB will move aft with ship design speed and Froude number. At low Froude number, the bow can be fairly blunt with cylindrical or elliptical bows utilized on slow vessels. On these vessels it is necessary to fair the stern to achieve effective flow into the propeller, so the run is more tapered (horizontally or vertically in a buttock flow stern) than the bow resulting in an LCB which is forward of amidships. As the vessel becomes faster for its length, the bow must be faired to achieve acceptable wave resistance, resulting in a movement of the LCB aft through amidships. At even higher speeds the bow must be faired even more resulting in an LCB aft of amidships.

Harvald [1983]

$$LCB = 9.70 - 45.F_n + (-0.8)$$

Schneekluth and Bertram [1998-second edition]

$$LCB = 8.80 - 38.9.F_n$$

$$LCB = -13.5 + 19.4.C_P$$

Here LCB is estimated as percentage of length, positive forward of amidships.

Summary

General Bulk cargo ships have made up a vast majority of merchant marine fleets due to their versatility and robust construction. With expected service life of 40 years many of the current vessels on the seas today are older than their crews.

Chapter 3

Container Ships

3.1. Introduction

Container ships and their containers have revolutionized the shipping business. Since the inception of containers back in 1956, this branch of the maritime business has grown phenomenally, becoming an integral part of the door-to-door international movement of goods and products. Making just a short journey by road, you are certain to pass a container travelling from or on its way to a container depot, and that will be just one of the millions of containers that are handled by ports and terminals around the world every year.

Container ships are designed to fit containers with standard measurements, with each one slotting into guides on the ship like pieces of a giant jigsaw puzzle. The carrying capacity of container ships, colloquially known as box ships, is measured in twenty-foot equivalent units, or TEU. Simply, this is the number of 20-foot containers that a ship can carry. An example of the container ship of MAERSK LINE is shown in Fig. 3.1.

3.2. Specifications of the ship

These ships have been steadily growing in size and today the largest ships afloat can carry 14,500 TEU, or 14,500 20-foot containers. Known as a Suezmax ship, this ship represents the maximum dimensions for Suez Canal transit. The next stage in container ship design will be the Malaccamax ship, capable of carrying 18,000 TEU and the maximum size for the Malacca Straits.

Figure 3.1: MSC Zoe [Freight Hub.2015]

The main characteristic of a container ship is that it depends on shore based lift-on/lift-off equipment, mainly container gantry cranes (also called portainers), to handle its cargo. The earliest purpose-built container ships in the 1960s were all gearless, i.e. without any shipboard cranes. Since then, the percentage of geared new builds has been decreasing overall, with only 7.5% of the container ship fleet capacity in 2009 being equipped with cranes. The introduction and improvement of shore side cranes have been a key to the success of the container ship.

Efficiency has always been key in the design of container ships. While containers may be carried on conventional break-bulk ships, cargo holds for dedicated container ships are specially constructed to speed loading and unloading, and to efficiently keep containers secure while at sea. A key aspect of container ship specialization is the design of the hatches. The hatch openings stretch the entire breadth of the cargo holds, and are surrounded by a raised steel structure known as the hatch coaming. On top of the hatch coamings are the hatch covers. Until the 1950s, hatches were typically secured with wooden boards and tarpaulins held down with battens. Today, some hatch covers can be solid metal plates that are lifted on and off the ship by cranes, while others are articulated mechanisms that are

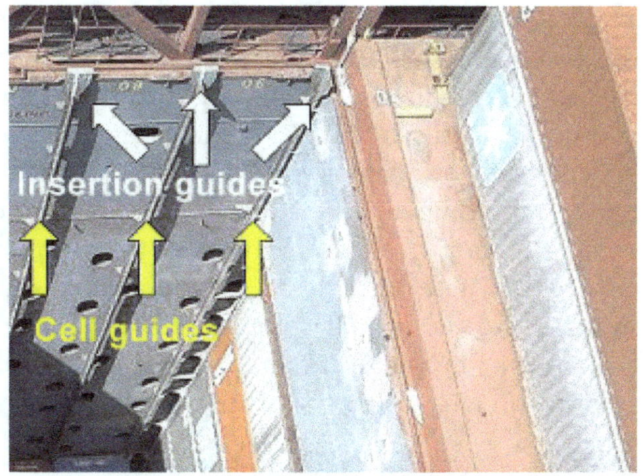

Figure 3.2:   Cell guides [Container handbook.2014]

opened and closed using powerful hydraulic rams. Another key component of dedicated container ship design is the use of cell guides. Cell guides, Fig. 3.2, are strong vertical structures constructed of metal installed into a ship's cargo holds. These structures guide containers into well-defined rows during the loading process and provide some support for containers against the ship's rolling at sea. So fundamental to container ship design are cell guides that organizations such as the United Nations Conference on Trade and Development use their presence to distinguish dedicated container ships from general break-bulk cargo ships.

Each container vessel is split into compartments, which are termed as bays, and depending on the size of the ship it will proceed from 01 to 40 (for example) where Bay 01 is the bay towards the Bow (the front) of the ship and Bay 40 is the Stern (the back) of the ship. Bays are numbered lengthwise from bow to stern with odd numbers for 20′ containers and even numbers for 40′ containers. The even number between two 20′ containers is used to define 40′ bays. The bay spaces for 20′ containers are numbered throughout fore to aft in this case 01, 03, 05 and so on up to 75. The bay spaces for 40′ containers are numbered throughout with even numbers: 02, 04, 06 and so on up to 74. A system of three dimensions is used in cargo plans to describe the position of a container aboard the ship. The first coordinate is the row, which starts at the front of the ship and increases aft. The second coordinate is tier, with the first tier at the bottom of the cargo holds, the second tier on top of that, and so forth. The third coordinate is the slot. Slots on the starboard side are given odd numbers and those on the port side are given even numbers. The slots nearest the centerline are given low numbers, and the numbers increase for slots further from the centerline. Container ships only take 20's, 40's, and 45 foot containers. 45 footers only fit above deck. 40 foot containers are the primary container size making up about 90% of all container shipping and since container shipping moves 90% of the world's freight, over 80% of the world's freight moves via 40 foot containers.

3.2.1. *Midship section of container ships*

For a typical container ship design, the midship section is shown in Fig. 3.3. It is however important, to understand the drawing from a designer's point of view. Some common features of the midship section of a container ship are discussed below:

- All container ships are double bottomed, to allow the double bottom spaces to be used as tanks.
- Container ships are also longitudinally framed, because the variable loading conditions often result in large hogging and sagging moments, which result in high longitudinal bending stresses.
- The shape of the midship section is almost box-like. From a technical perspective, it has high midship area coefficient, ranging from 0.75 to 0.85
- The bilge strake is the angular plate that joins the inner side shell and the tank top plating. Since the presence of this plate would prevent the stowage of containers at the corner of the section, the length of this strake is kept to a minimum. In most recent cases, however, container ships are not provided with bilge strakes at all, in order to ensure maximum stowage capacity.
- The most important structural feature of a container ship is the torsion box, which we will discuss in detail in one of the following sections.

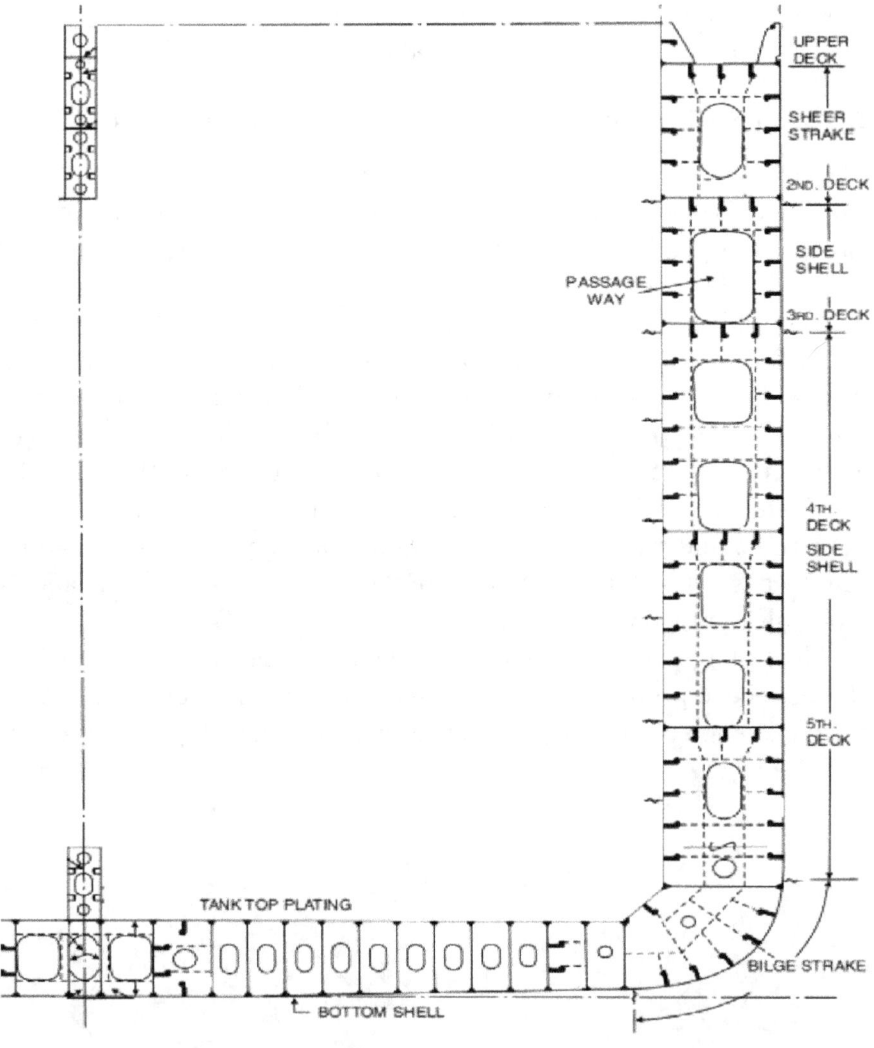

Figure 3.3: Midship section of a container ship [Marine Insight.2015]

- Container ships are usually equipped with no hatches. That is, the ship has no continuous main deck running full breadth all along the ship. This open box-like structure (relate with the midship section figure), enables easy stowage of containers from the tank top to the highest level above main deck level. The only decks are within the double hull, which are more like stringers running full length, and provide passageway along the length of the ship.

3.2.2. *Torsion box in double hulled container ships*

The entire structure of an open hatch has a very low polar moment of inertia. This results in a very low torsional strength. Thus, in order to strengthen open box like structure against torsion, additional material is added to the rim. In other words, one simply increases the polar moment of inertia of the cross section of the structure. We know that the hull of a container ship is an open box like structure. When a ship is at sea, it is subjected to various wave loads. In one of the load cases, when the direction of the waves is at approximately 45 degrees to the velocity of the ship, port side of the forward section and the starboard side of the aft section would experience a wave crest at the same time, and vice versa. This results in a type of periodic loading which causes the hull to twist. This effect is called torsion, and is shown in the following Fig. 3.4.

The lack of support and strengthened sections of a hatch opening of a vessel can produce mild effects on ship structure. The same effect when extrapolated to the grand scale of a container ship, would result

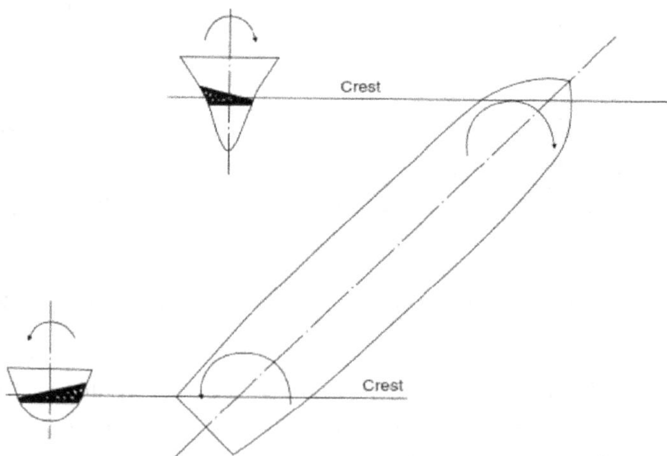

Figure 3.4: Torsion in a container ship moving in quarter seas [Marine Insight.1993]

in devastating failures of the hull structure due to torsion. In order to prevent this, the topmost edges of the port and starboard sides of a container ship are strengthened with high scantling web sections, creating a box like structure at every frame. This is called torsion box. A torsion box runs along the entire length of the ship from the aft peak bulkhead to the forward collision bulkhead.

The following Fig. 3.5 shows the torsion box of the double hull container ship. Note that the width of the web plate in the torsion box is higher than the web plate used around the passageway below it. In addition, the webs at passageways are at a spacing of three to four frame spaces, but the webs of a torsion box would be present at every frame.

Classification societies have laid down separate set of rules for the design of torsion box for container ships. Today, the hull girder is modelled on an FEM platform and its torsional response is analyzed

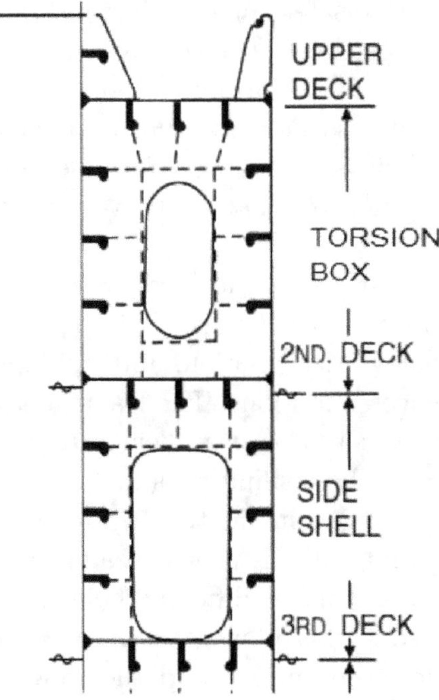

Figure 3.5: Torsion box of a container ship [Marine Insight.1993]

for different sea states. The torsion box needs to be redesigned if the torsional stresses on the hull girder are above the safe limits. The inspection of torsion box is always treated as a high priority in annual structural surveys, as it plays a significant role in determining the strength of the hull structure.

3.2.3. *Stowage of containers*

The stowage of containers on a container ship is another aspect that the designer has to deal with. Though it may come across as something insignificant, improper stowage has resulted in most of the accidents related to container ships.

- Containers are always stowed with the longer dimension along forward to aft. This is because; the ship is more prone to rolling motions than pitching or yawing. Stowage of containers in this orientation ensures less space for the cargo to shift within the container, ensuring more safety against impact damage of the cargo.
- Below the uppermost deck, the containers are restrained against lateral or longitudinal motion by cell guides. These are basically angle sections that also help as guides for containers when they are loaded onto the ship. However, these do not form a part of the primary structure, that is, they do not take up the hull stresses.
- Above the uppermost deck, containers are stowed and their motion is restricted by means of lashings. Twist locks fitted between the containers prevent vertical motion, and lashing prevent the longitudinal and transverse motions. The lashings are usually deployed from lashing bridges that are at height intervals of one or two tiers of containers. The lashing rods are secured at their ends by turnbuckles which maintain the tension in the lashings.
- The container loading plan is provided along with the design, and it specifies the positions of different containers on the ship, at different load cases. This plan takes into consideration the fact that the number of containers and the weight of cargo in each would differ on each voyage. And the stowage would also have to take into consideration, the port at which each container has to be

unloaded. Heavier containers cannot be stowed above the lighter ones, as it would raise the centre of gravity of the vessel, reducing the stability margin.

This complicity of container ship design is therefore solved by means of specially designed computer programs to generate container loading plans for a every particular loading case, which factors in, the series of ports a vessel needs to call, and also the strength and stability aspects of the ship. Another critical factor is, the visibility from the bridge. The containers loaded above the deck and forward of the navigation bridge are to be loaded such that the line of sight from the bridge is not affected. That is why, the stack of containers forward of the bridge reduces in height as one moves to the forward-most stack. This, however, reduces the total amount of containers that can be carried by the ship. Hence, many ultra large container ships (e.g. Maersk Triple E class) have their superstructures shifted to the midship, in order to be able to accommodate containers to full height aft of the superstructure.

3.3. Important parameters

General trends in the dimensions of container ships are illustrated in Tables 3.1 and 3.2:

Table 3.1: Dimensions of container ships [Marine Notes]

Name	Capacity (TEU)	Length (m)	Beam (m)	Draft (m)
Ultra Large Container Vessel (ULCV)	14,501 and higher	366	49	15.2 m (49.9 feet) and deeper
New Panamax	10,000–14,500	366	49 m (160.7 feet)	15.2 m (49.9 feet)
Post Panamax	5,101–10,000	366	49 m (160.7 feet)	15.2 m (49.9 feet)
Panamax	3,001–5,100	294.13	32.31 m (106 feet)	12.04 m (39.5 feet)

Table 5 gives some of the parameters of a container ship.

Table 3.2: Hydrostatics of a container ship [Dr Prasanta Sahoo, "Preliminary Ship Design", Internet data]

Parameter	Value
C_B	0.5–0.7
C_P	0.60–0.68
C_M	0.97–0.98
Velocity Range	14–26 knots
B/T	2.25<=B/T<=3.75 (conventional monohull)
	~5 (heavily draft limited vessels)
	B/T<=[9.625-7.5C_B] (for ensuring proper flow on propeller)
T/D	0.7 for type B freeboard
	0.7–0.8 for B-60 freeboard
L/B	6.5 for L>=130 m
	4.0 + 0.025 (L − 30) for 30 m<=L<=130 m

The draft is generally given by $T = 4.536 [dwt/1000]^{0.290}$ m.

3.4. Initial estimate of stability

3.4.1. *Vertical centre of buoyancy KB*

For a container ship C_M is > 0.9, hence according to Posdumine and Lackenby,

$$KB/T = (1 + C_{VP})^{-1}.$$

Regression formulations are as follows:

$$KB/T = 0.90 - 0.36\,C_M$$

$$KB/T = (0.90 - 0.30\,C_M - 0.10\,C_B)$$

$$KB/T = 0.78 - 0.285\,C_{VP}$$

Where KB: Vertical centre of buoyancy and C_{VP}: Vertical Prismatic Coefficient

3.4.2. *Metacentric radius:* $\mathbf{BM_T}$ *and* $\mathbf{BM_L}$

Moment of Inertia coefficient C_I and C_{IL} are defined as

$$C_I = I_T/LB^3,$$

$$C_{TL} = I_L/L^3B.$$

3.4.3. *Transverse stability*

$KG/D = 0.63$ to 0.70 for container ships

$$KM_T = KB + BM_T$$

$$GM_T = KM_T - KG$$

Correction for free surface must be applied over this. Then,

$$GM'_T = GM_T - 0.03\,KG \text{ (assumed)}.$$

This GM'_T should satisfy IMO requirements.

3.4.4. *Longitudinal stability*

$GM_L \sim BM_L = I_L/\nabla = (C_{IL}.L^3.B)/L.B.T.C_B = C_{IL}.L^2/T.C_B$
$MCT\ 1\,cm = \nabla.GM/100.L_{BP} = L.B.T.C_B.C_{IL}.L^2/100.T.C_B.$
$L = C_{IL}.L^2.B/100$

3.4.5. *Longitudinal centre of buoyancy*

It follows the same trends as mentioned previously. The following two formulae are important in this respect.

Harvald [1983]

$$LCB = 9.70 - 45.F_n + (-0.8)$$

Schneekluth and Bertram [1998-second edition]

$$LCB = 8.80 - 38.9.F_n$$

$$LCB = -13.5 + 19.4.C_P$$

Here LCB is estimated as percentage of length, positive forward of amidships.

3.5. Some equations for container ships

The following equations have been taken from "Sahoo Prasanta, Computer aided optimization of the midship section of container ships based on longitudinal strength using non-linear optimization methods".

These equations give the link between the various parameters namely L_{PP}, TEU, Speed, Depth and Breadth of a container ship.

For $V_S = 15$ to 21 knots (First Generation Ship):

$$L_{PP} = 8.3064.V_S^{0.454}.TEU^{0.233}$$

$$B = 0.55244.V_S^{0.3552}.TEU^{0.4101}$$

$$D = 0.35477.V_S^{1.02167}.TEU^{0.44134}$$

$$T = 0.6008.V_S^{0.275}.TEU^{0.283}$$

For $V_S = 16$ to 28 knots (Second Generation Ship):

$$L_{PP} = 5.8422.V_S^{0.5537}.TEU^{0.2453}$$

$$B = 2.862.V_S^{0.1886}.TEU^{0.24183}$$

$$D = 0.784234.V_S^{0.36161}.TEU^{0.272965}$$

$$T = 1.7252.V_S^{0.16996}.TEU^{0.172922}$$

For $V_S = 16$ to 28 knots (Third Generation Ship):

$$L_{PP} = 6.0052.V_S^{0.551435}.TEU^{0.25426}$$

$$D = 0.7804.V_S^{0.30065}.TEU^{0.31222}$$

$$T = 1.614.V_S^{0.14388}.TEU^{0.19782}$$

where L_{PP}: Length between perpendiculars, V_S: Speed of the vessel, D: Depth, T: Draft.

Chapter 4

Bulk Carrier

4.1. Introduction

Bulk carriers are a type of ship, which transports cargoes in bulk quantities. The cargo in bulk ships is considered loose cargo i.e. without any specific packaging to it and generally contains items like food grains, ores and coals and even cement. Bulk carriers are often referred to as the workhorses of the maritime business bred specifically to carry dry cargoes such as those mentioned above.

Since their inception towards the mid-19th century, bulk vessels have been revolutionized and streamlined in order to facilitate greater ease for their owners and operators, presently. First conceived in the 1950's, purpose built Bulk carriers filled a vital role in the post-war shipping industry. A majority of the allied merchant fleet at the time comprised of liberty ships some 2,400 vessels. The Liberty ships were designed and built with expediency and minimal cost in mind. Using outdated but reliable coal engines and equipped with 5 cargo holds equipped to handle palletized cargo. However, the bulk cargo consisting of ore, rock, coal and grains proved to be difficult. The 'tweendeckers that allowed the liberty ships to carry more pallets made stowing and discharging the loose coal or ore a major nuisance. Shipping grain however was another matter entirely. The complex process required to ship grain started with construction and installation of grain fittings or "shifting boards" a series of plywood walls and beams to prevent the gain from shifting mid transport. These fittings could cost several days to fit and a team of skilled carpenters costing valuable time and money.

31

Figure 4.1: General bulk carrier [Marine Insight.2016]

The solution to these inefficiencies were first build in 1954, the OS-type became the precursor to the modern bulk carrier. Featuring the sloped wind holds which prevent shifting of loose cargo and large spacious hatches reduced transport cost and turn around time dramatically. Shore facilities were required to unload these carriers as most ships were fitted without gear. A crane with grab attachment and small bulldozers is the primary method used for unloading these vessels even to this day.

Bulk Carriers have one more stage of evolution to undergo before reaching the modern era. The self-unloading vessels build in 1956 have several interesting design characteristics. The large sloped cargo tanks bottom out into hatches opening onto a conveyor belt. The belt then carries the cargo to the bow of the vessel where a boom conveyor transfers the cargo onshore. This process has completely revolutionized how bulk carriers were viewed in the shipping industries. However, there are numerous drawbacks to this design. Each vessel cost $\sim$ 25% more to build than a standard vessel and requires near constant maintenance on the conveyor systems to maintain operation status.

Figure 4.2: Representation of self-unloading bulk carrier [Bulk Carrier Guide.2010]

The largest issue facing the self-unloading design is the reduction in cubic volume of hold capacity. The reduction compared to vessels of a similar size presents a challenge to the shipping company and more expense per tonn on the consumer. The saving grace of the design is in the transportation of dense material. The bulk carrier fleet centered in the American Great Lakes region is made up nearly entirely of these self-unloading bulk carriers.

In addition to carrying dry cargo like the ones specified above, a bulker is also engaged at times to carry liquefied cargoes. The liquefied cargo carried may include oil, petrol and various other liquid chemical substances.

Although bulk carriers have been employed since the 1850s, their appropriate definition and interpretation can be found in the SOLAS Convention — year 1999. However, over the years various other interpretations have also been added to the official definition, which are now being employed quite effectively.

Apart from the major classifications based on carrying capacity, there are several other classifications applicable to certain specific navigation channels. These navigational channel classifications however do not form a part of the international shipping domain but are restricted to certain geographical shipping arenas.

In order to bring about a better quality to the cargo ship, it has been proposed that these vessels be built according to Common Structural Rules or CSR. Vessels that are built according to the CSR

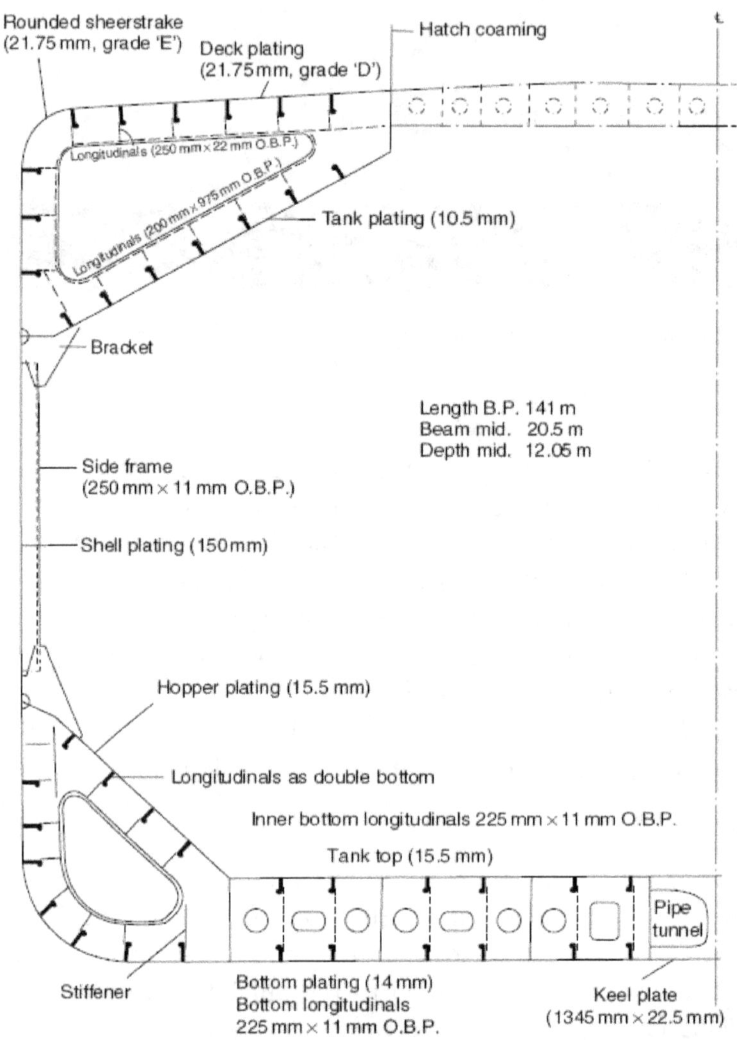

Figure 4.3: Midship section of bulk carrier [Learn Ship Design (LSD)]

specifications are annotated with the CSR mark, which helps to boost the vessel's credibility in the global sector.

Recognizable by the large box-like hatches on deck, which slide out for loading, bulk carriers vary enormously in size. From tiny single-hold coasters with a 500 ton load, to huge Capesize carriers

Table 4.1: Size of different types of bulk carriers [Wikipedia]

Bulk Vessel Category	Size in DWT
Handysize	10,000 to 35,000
Handymax	35,000 to 59,000
Panamax	60,000 to 80,000
Capesize	80,000 and over

of up to 230,000 tons, which might take on the task of transporting cargoes of iron ore or coal from Australian mines to Chinese steel mills.

Known for their efficiency and cost benefit, some 6,000 bulk carriers, or bulkers as they are sometimes referred to, trade around the world. Some of the more sophisticated bulk carriers have their own cargo gear on board, in the form of derricks, cranes or conveyor systems, while other make use of specialist port equipment to blow rab, pump or convey the cargo away from the ship and into customized storage buildings.

4.2. Important parameters

Bulkers are segregated into six major size categories: small, Handysize, Handymax, Panamax, Capesize, and very large. Very large bulk and ore carriers fall into the Capesize category but are often considered separately.

The draft for a bulk carrier is given by $T = 0.66D + 0.9m$.

4.3. Initial estimate of stability

4.3.1. *Vertical centre of buoyancy KB*

For a bulk carrier C_M is > 0.9, hence according to Posdumine and Lackenby,

$$\mathrm{KB}/\mathrm{T} = (1 + C_{\mathrm{VP}})^{-1}.$$

Table 4.2: Hydrostatics of a bulk carrier [Dr Prasanta Sahoo, "Preliminary ship design", Internet data]

Parameter	Value
L	Up to 365 m
C_B	0.8–0.85
C_P	0.76–0.85
C_M	0.97–0.98
Velocity Range	12–17 knots
B/T	$2.25 <= B/T <= 3.75$ (conventional monohull)
	~ 5 (heavily draft limited vessels)
	$PB/T <= [9.625{-}7.5C_B]$
	(for ensuring proper flow on propeller)
T/D	0.7 for type B freeboard
	0.7–0.8 for B-60 freeboard

Regression formulations are as follows:

$$KB/T = 0.90 - 0.36C_M$$

$$KB/T = (0.90 - 0.30C_M - 0.10C_B)$$

$$KB/T = 0.78 - 0.285C_{VP}$$

Where KB: Vertical centre of buoyancy and C_{VP}: Vertical Prismatic Coefficient

4.3.2. *Metacentric radius: BM_T and BM_L*

Moment of Inertia coefficient C_I and C_{IL} are defined as

$$C_I = I_T/LB^3,$$

$$C_{TL} = I_L/L^3B.$$

4.3.3. *Transverse stability*

$$KG/D = 0.63 \text{ to } 0.70 \text{ for cargo carrying ships}$$

$$KM_T = KB + BM_T$$

$$GM_T = KM_T - KG$$

Correction for free surface must be applied over this. Then,

$$GM'_T = GM_T - 0.03 \text{ KG (assumed)}.$$

This GM'$_T$ should satisfy IMO requirements.

4.3.4. *Longitudinal stability*

$GM_L \sim BM_L = I_L/\nabla = (C_{IL} \cdot L^3 \cdot B)/L \cdot B \cdot T \cdot C_B = C_{IL} \cdot L^2/T \cdot C_B$
$MCT1cm = \nabla \cdot GM/100 \cdot L_{BP} = L \cdot B \cdot T \cdot C_B \cdot C_{IL} \cdot L^2/100 \cdot T \cdot C_B \cdot L = C_{IL} \cdot L^2 \cdot B/100$

4.3.5. *Longitudinal centre of buoyancy*

Follows the same trends as mentioned previously. The following two formulae are important in this respect.

Harvald [1983]

$$LCB = 9.70 - 45.F_n + (-0.8)$$

Schneekluth and Bertram [1998-second edition]

$$LCB = 8.80 - 38.9.F_n$$

$$LCB = -13.5 + 19.4.C_P$$

Here LCB is estimated as percentage of length, positive forward of amidships.

Chapter 5

Tankers

5.1. Introduction

Tanker family of ships include some of the world's largest mechanically propelled floating objects. Carrying a wide range of fluid cargoes from gas and crude oil to alcohol and acids, tankers form a indispensable part of the global economy.

Crude oil tankers, the largest of the family, carry oil from the production facilities, such as in the Middle East Gulf, to the refiners, such as North Europe. Product tankers are the smaller sisters of the tanker family, carrying refined products, such as gasoline or diesel to distribution hubs. Other important members of the tanker family include chemical tankers, which are specially designed to carry corrosive acids and wines among other things, and gas carriers, moving methane and other gases in a liquefied or compressed state.

What makes the tanker fleet particularly special is its adaptability. Changing demands for its cargoes mean that energy transportation business can dramatically shift from one side of the world to another. Its requirement for responsive nature means that ships must be able to be switched between geographical areas to add or reduce capacity as the need arises.

5.2. Crude oil tankers

Crude oil tankers are designed to carry crude oil from oilfields to refineries around the world usually making the return journeys full of ballast. The size of these ships increased steadily, peaking in the 1970s in order to bring the cost of oil transportation to the absolute

Figure 5.1:   Arctic princess [Seanews 2013]

minimum. The era of the 'supertanker', which commenced in the 1960s, quickly led to the construction of a very large and ultra-large crude carriers (VLCCs and ULCCs) which are the largest ships ever built.

As cargo-handling technology progressed, many tankers were generally fitted with steam coil heating in the cargo tanks to keep heavier grades of oil viscous and speed up the discharge of cargo. They were also equipped with crude oil washing equipment for tank cleaning and inert gas systems to reduce the risks of fire and explosion.

Some of the latest tanker new buildings incorporate other extensive safety features from double hulls, double engine rooms, propulsion and steering equipment. Ship navigational and communications systems have also become increasingly sophisticated in recent years.

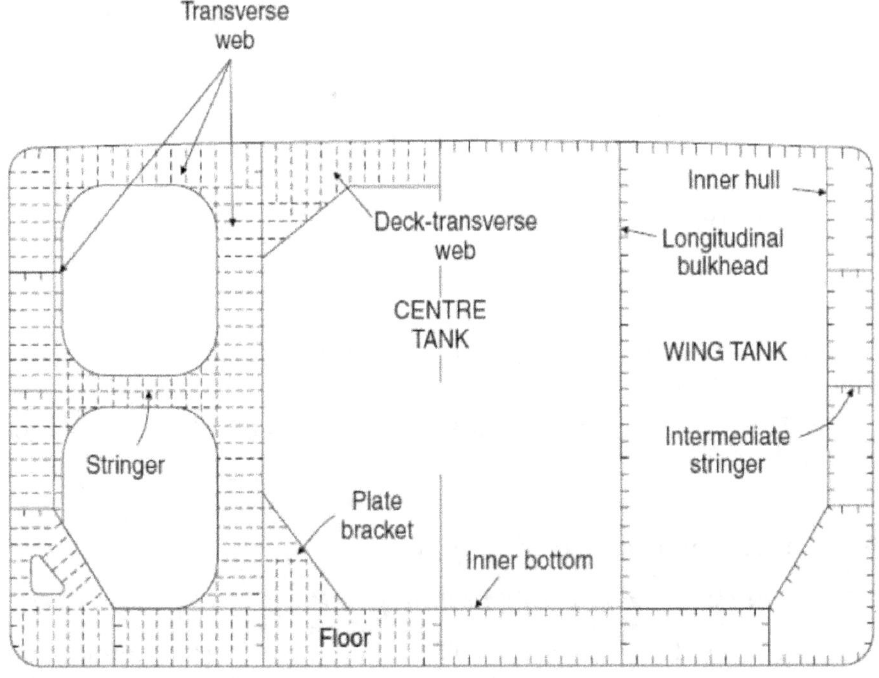

Figure 5.2: Midship section of double hull oil tanker [Marine Insight]

Table 5.1: Hydrostatics of an oil tanker [Dr Prasanta Sahoo, "Preliminary Ship Design", Internet data]

Parameter	Value
L	Up to 365 m
C_B	0.8–0.85
C_P	0.82–0.90
C_M	0.98–0.99
Velocity Range	12–16 knots
B/T	2.25 <= B/T <= 3.75 (conventional monohull)
	~5 (heavily draft limited vessels)
	B/T <= [9.625–7.5C_B] (for ensuring proper flow on propeller)
T/D	0.8 for type A freeboard (tankers)
	(T/D <0.8 for double hull tankers)
L/B	6.5 for L >= 130.0 m
	4.0 + 0.025 (L−30) for 30 m <= L <= 130 m

Table 5.2: Dimensions of different types of oil tankers [Google search]

Tanker category	Length	Breadth	Draft
Coastal Tanker	205 m	~24 m	~7.9 m
Aframax	245 m	44.2 m	11.6 m
Suezmax	285 m	50 m	16 m
VLCC	330 m	60 m	20 m
ULCC	415 m	—	—

The expression for draft of a tanker is given by T=4.536 $[\mathrm{dwt}/1000]^{0.290}$ m.

Table 5.2 gives the trends in the dimensions of different types of tankers.

5.3. Initial estimate of stability

5.3.1. *Vertical centre of buoyancy KB*

For a tanker C_M is >0.9, hence according to Posdumine and Lackenby,

$$KB/T = (1 + C_{VP})^{-1}.$$

Regression formulations are as follows:

$$KB/T = 0.90 - 0.36C_M$$

$$KB/T = (0.90 - 0.30C_M - 0.10C_B)$$

$$KB/T = 0.78 - 0.285C_{VP}$$

Where KB: Vertical centre of buoyancy and C_{VP}: Vertical Prismatic Coefficient.

5.3.2. *Metacentric radius: BM_T and BM_L*

Moment of Inertia coefficient C_I and C_{IL} are defined as

$$C_I = I_T/LB^3,$$

$$C_{TL} = I_L/L^3B.$$

5.3.3. *Transverse stability*

$KG/D = 0.63$ to 0.70 for container ships

$$KM_T = KB + BM_T$$

$$GM_T = KM_T - KG$$

Correction for free surface must be applied over this. Then,

$$GM'_T = GM_T - 0.03 \, KG \text{ (assumed)}.$$

This GM'$_T$ should satisfy IMO requirements.

5.3.4. *Longitudinal stability*

$GM_L \sim BM_L = I_L/\nabla = (C_{IL}.L^3.B)/L.B.T.C_B = C_{IL}.L^2/T.C_B$

$MCT \, 1 \, cm = \nabla.GM/100.L_{BP} = L.B.T.C_B.C_{IL}.L^2/100.T.C_B.$

$L = C_{IL}.L^2.B/100$

5.3.5. *Longitudinal centre of buoyancy*

It follows the same trends as mentioned previously. The following two formulae are important in this respect.

Harvald [1983]

$$LCB = 9.70 - 45.F_n + (-0.8).$$

Schneekluth and Bertram [1998-second edition]

$$LCB = 8.80 - 38.9.F_n.$$

$$LCB = -13.5 + 19.4.C_P.$$

Here LCB is estimated as percentage of length, positive forward of amidships.

Chapter 6

Offshore Supply Vessels

6.1. Introduction

Offshore supply vessels are regularly used to transport goods, supplies or equipment in support of exploration or production of offshore mineral or energy resources. Offshore supply vessels are typically operated by ship owners: either companies set up specifically to own and operate such vessels or companies combine with other vessel operations like salvage, shipping, etc. These vessels carry out different operations necessary for floating drilling rigs, as well as moored or fixed production platforms.

The overarching offshore support vessels (OSVs) classification can be divided into a number of types according to the operations they perform: seismic survey ships, platform supply vessels (PSV), anchor handling tugs, anchor handling tug and supply vessels (AHTS), offshore construction vessels (OCV), ROV support vessels, dive support vessels, stand-by vessels, inspection, maintenance and repair vessels (IMR) and variety of combinations of these.

6.2. Mission & design constraints

The general mission of supporting offshore oil and gas exploration and production can be subdivided according to specific mission requirements. Also vessels may be involve in more than one kind of such specific missions. Like every other design criteria, optimizing based on priority is done in the design process, with a concentration on the safety requirements.

Figure 6.1:   Offshore supply vessel [ExxonMobil.2019]

These specific missions include:

- seismic survey to locate oil and gas bearing areas
- towing of rigs and platforms to their location, positioning them, and laying anchoring & mooring equipments
- supplying rigs and platforms with necessary personnel, equipment, stores, provisions, etc.
- Subsea operations like ROV operation, diving support, inspection, and maintenance
- safety standby

Following are the different types of offshore supply vessels:

6.2.1. *Seismic survey ship*

A vessel that maps out geological structures in the seabed by firing air guns, transmitting sound waves into the bottom of the sea. The echo of the shot is captured by hydrophones that are towed behind the vessel. Dedicated seismic survey vessels are highly specialized ships. The working decks are enclosed but typically are open at the stern, lower levels typically have the air gun handling system and storage, and at higher levels, they have winches and storage reels for streamers. The ship itself must be capable of accurate track, station

Figure 6.2:   Ramform Viking [PGS.2019]

keeping and the propulsion system must have low radiated noise, and minimal propeller induced noise to avoid interference with the survey equipment.

6.2.2. *Platform Supply Vessel (PSV)*

The PSV are designed for supplying offshore drilling rigs and production platforms with necessary equipment, stores and drilling consumables. The cargo typically consists of cement, baryte and bentonite transported as dry powders; drill water; oil or water based liquid mud, methanol and chemicals for specialized operations. The PSV loads at a shore base. Liquid cargo is carried in double bottom tanks, dry bulk cargoes in special pneumatic pressure tanks, and equipment and drill pipes on the aft open deck. At the rig or platform, the liquid and powder cargoes are pumped up or transferred pneumatically while deck cargo is handled by the rig crane.

A typical PSV operating profile shows the vessel spending about 25% of the service in harbor loading and unloading, 40% sailing under load with 14–16 knot range and 35% loading or discharging at sea, often in strong winds, high seas and strong currents.

Figure 6.3: Daleel platform supply vessel [Daleel.2019]

PSVs are used for transporting supplies to the rigs and oil platform and return other cargoes to shore. The supplies include fuel, fresh water, equipment's, consumables, stores and provision for the operation of the oil platforms and for the personnel onboard. The design and construction of these vessels depend on the working environment — weather conditions, distance from the shore, and hence design varies by location.

6.2.3. *Anchor Handling Tug Supply Vessel (AHTS)*

The AHTS combines a number of functions in a single hull. These include handling the anchors and mooring chains for drilling rigs, towing of rigs and platforms together with subsequent positioning on site, and platform supply duties. The required bollard pull has a powerful influence on the design, since this defines the power need, the propeller size, hull shape and depth aft to give the necessary propeller immersion. Hull beam and shape shall give good stability, particularly when heavy moorings and anchors are suspended from the stern. Anchor handling requires high power, winch capacity, deck space aft, storage bins for rig chains and auxiliary handling

Figure 6.4: Farstad shipping AHTS [Marine Link.2019]

equipment. A stern roller is used to ease the passage of wires and anchors over the stern of the vessel during deploying or weighing of the anchor.

The anchor handling tug is a naval vessel that is solely concerned with the objective of either tugging or towing an oil-rig or a ship. When it comes to oil rigs, these tugs form the most important necessity as without their help, it would be impossible to place oil rigs in the required sea and oceanic areas.

6.2.4. *Design aspects and additional features*

Anchor handling tug vessels or systems have a crane like equipment (known as the winch) that can be attached to the oil rigs and then propelled forth in the water. The "anchor supply", mentioned as a part of the vessel's name, is then allowed to be sunk into the seawater in order to keep the rigs steady.

6.2.5. *Construction support vessels*

Dynamically-positioned Class 3 vessels with large unobstructed deck areas, substantial accommodation capacity and significant surface and subsea heavy lift crane capability, able to support surface and subsea construction and installation projects, as well as inspection,

Figure 6.5: Normand baltic CSV [AMEMaritime.2017]

repair and maintenance (IRM) programs. Construction Support Vessels are designed to provide tailored solutions and facilitate larger projects that often require such vessels to remain on location for long periods of time. These hardy ships are designed to operate 24/7 in some of the harshest seas imaginable. With between 100–200 project crew aboard, offshore construction vessels are often the on station for months at a time providing assistance to the rig construction teams. These vessels are often equipped with safety gear and firefighting capabilities to provide assistance in any situation. Built significantly larger and more sophisticated that other members of the offshore vessel family these vessels are often the first on and last to leave a job site.

6.2.6. *Diving Support Vessel (DSV)*

A vessel provided with diving equipment and used for underwater work such as the maintenance and inspection of mobile platforms, pipelines and their connections, wellheads, etc. The DSV is fitted with a moon pool — a hole in the middle of the vessel open to the sea — through which divers, remotely-operated vehicles and other equipment is passed to and from the worksite. The DSV maintain an almost exact position over the worksite during diving operations.

Figure 6.6:   BOURBON evolution dive support vessel [BOURBON.2017]

In order to accomplish this, it utilizes satellite positioning and a DP3 class positioning system to maintain optimal location in any conditions.

6.2.7. *Inspection Maintenance and Repair (IMR) vessel*

A dynamically positioned ship-shaped offshore unit provided with equipment for well stimulation or maintenance (e.g. coil tubing). Such vessels are often able to carry out other tasks, such as ROV operations and general supply duties.

6.2.8. *ROV support vessels*

Dynamically Positioned (DP) Vessels from which ROV operations are conducted. ROV Support Vessels are equipped with computer-controlled, precision, position-keeping capabilities with added redundancy features, such as multiple computers, thrusters and reference systems. Such vessels have additional cabins, mess room facilities and Client offices, to comfortably accommodate the Client's ROV support crews. Like the Diving support vessel, this classification typically is equipped with a moon pool. At minimum an aft A-frame or winch gear is often quipped.

Chapter 7

Tugs

7.1. Introduction

A **tug** (**tugboat**) is a boat or ship that maneuvers vessels by pushing or towing them. Tugs move vessels that either should not move by themselves, such as ships in a crowded harbor or a narrow canal, or those that could not move by themselves, such as barges, disabled ships, log rafts, or oil platforms. Tugboats are powerful for their size and strongly built, and some are ocean-going. Some tugboats serve as icebreakers or salvage boats. Early tugboats had steam engines, but today most have diesel engines. These are relatively smaller but very powerful for their size. In addition to these, tugboats are also used as icebreakers or salvage boats and as they are built with firefighting guns and monitors, they assist in the firefighting duties especially at harbors and when required even at sea.

7.1.1. *Types of tugboats*

7.1.1.1. *Deep-sea tugs*

Seagoing tugs (deep-sea tugs or ocean tugboats) fall into four basic categories:

- The standard seagoing tug with model bow that tows its "payload" on a hawser.
- The "notch tug" which can be secured in a notch at the stern of a specially designed barge, effectively making the combination a ship. This configuration is dangerous to use with a barge which is "in ballast" (no cargo) or in a head- or following sea. Therefore,

53

Figure 7.1:   Tugboat [Maritime-Executive.2018]

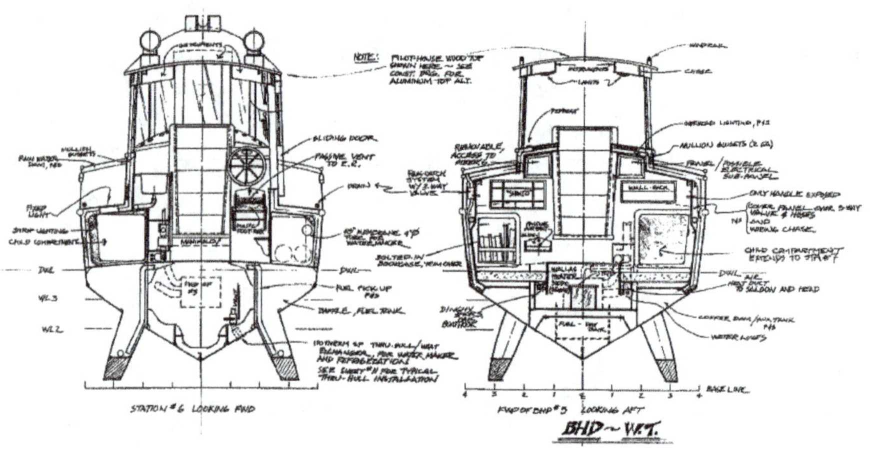

Figure 7.2:   Midship section of a tug [Kasten Marine Design]

"notch tugs" are usually built with a towing winch. With this configuration, the barge being pushed might approach the size of a small ship, with interaction of the water flow allowing a higher speed with a minimal increase in power required or fuel consumption.

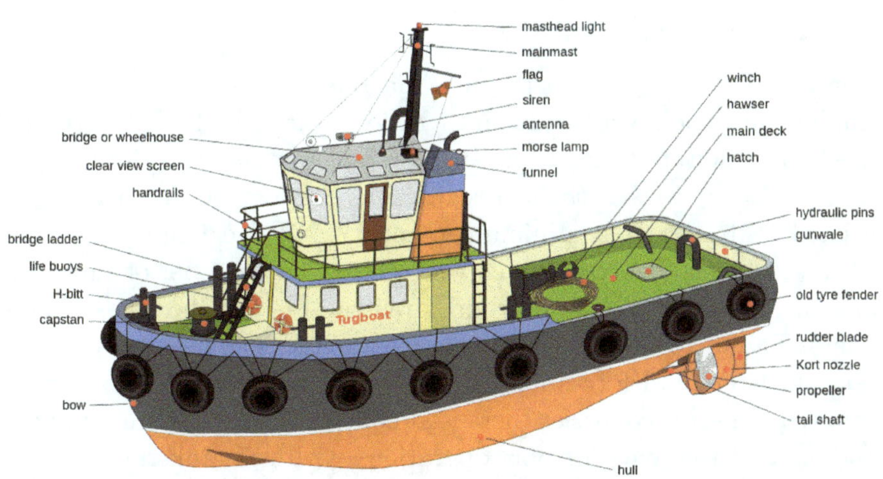

Figure 7.3: Image of a tugboat [Wikipedia]

- The "integral unit", or "integrated tug and barge" (ITB), comprises of specially designed vessels that lock together in such a rigid and strong method as to be certified as such by authorities (classification societies) such as the American Bureau of Shipping, Lloyd's Register of Shipping, Indian Register of Shipping, Det Norske Veritas or several others. These units stay combined under virtually any sea conditions and the tugs usually have poor sea-keeping designs for navigation without their barges attached. Vessels in this category are legally considered to be ships rather than tugboats and barges must be staffed accordingly. These vessels must show navigation lights compliant with those required of ships rather than those required of tugboats and vessels under tow.

- "Articulated tug and barge" (ATB) units also utilize mechanical means to connect to their barges. The tug slips into a notch in the stern and is attached by a hinged connection. ATBs generally utilize Intercon and Bludworth connecting systems. ATBs are generally staffed as a large tugboat, with between seven and nine crew members. The typical American ATB operating on the east coast customarily displays navigational lights of a towing vessel pushing ahead, as described in the 1972 ColRegs.

7.1.1.2. *Harbor tugboats*

Compared to seagoing tugboats, harbor tugboats are generally smaller and their width-to-length ratio is often higher, due to the need for a lower draught. In smaller harbors these are often also termed *lunch bucket boats*, because they are only manned when needed and only at a minimum (captain and deckhand), thus the crew will bring their own lunch with them. The number of tugboats in a harbor varies with the harbor infrastructure and the types of tugboats. Things to take into consideration includes ships with/without bow thrusters and forces like wind, current and waves and types of ship (e.g. in some countries there is a requirement for certain numbers and sizes of tugboats for port operations with gas tankers).

7.1.1.3. *River tugboats*

River tugs are also referred to as towboats or push boats. Their hull designs would make open ocean operation dangerous. River tugs usually do not have any significant hawser or winch. Their hulls feature a flat front or bow to line up with the rectangular stern of the barge, often with large pushing knees.

7.1.2. **Important parameters**

Table 7.1 gives some parameters for a tugboat.

Table 7.1: Hydrostatics of a Tugboat (modern American practice) [Argyriads Modern Tug Design]

Parameter	Ocean	Coastwise	Harbor
L/B	4.1	3.7	3.3
L/D	8.4	7.2	7.8
B/H	2.2	2.2	2.9
KG/D	0.89	0.91	0.88
C_B	0.52	0.48	0.50
C_P	0.65	0.66	0.64
C_M	0.80	0.80	0.78
GM light	1.70	3.30	2.0

7.1.3. *Initial estimate of stability*

7.1.3.1. *Vertical centre of buoyancy KB*

For a general cargo ship C_M is > 0.9, hence according to Posdumine and Lackenby,

$$KB/T = (1 + C_{VP})^{-1}.$$

Regression formulations are as follows:

$$KB/T = 0.90 - 0.36C_M$$

$$KB/T = (0.90 - 0.30C_M - 0.10C_B)$$

$$KB/T = 0.78 - 0.285C_{VP}$$

Where KB: Vertical centre of buoyancy and C_{VP}: Vertical Prismatic Coefficient

7.1.3.2. *Metacentric radius: BM_T and BM_L*

Moment of Inertia coefficient C_I and C_{IL} are defined as

$$C_I = I_T/LB^3,$$

$$C_{TL} = I_L/L^3B.$$

7.1.3.3. *Transverse stability*

KG / D = 0.90 for trawlers and tugs,

$$KM_T = KB + BM_T$$

$$GM_T = KM_T - KG$$

Correction for free surface must be applied over this. Then,

$$GM'_T = GM_T - 0.03\,KG \text{ (assumed)}.$$

This GM'_T should satisfy IMO requirements.

7.1.3.4. *Longitudinal stability*

$$GM_L \sim BM_L = I_L/\nabla = (C_{IL}.L^3.B)/L.B.T.C_B = C_{IL}.L^2/T.C_B$$

$$MCT\ 1\,cm = \nabla.GM/100.L_{BP} = L.B.T.C_B.C_{IL}.L^2/100$$
$$.T.C_B.L = C_{IL}.L^2.B/100$$

Longitudinal Centre of Buoyancy
It follows the same trends as mentioned previously. The following two formulae are important in this respect.

Harvald [1983]

$$LCB = 9.70 - 45.F_n + (-0.8).$$

Schneekluth and Bertram [1998–second edition]

$$LCB = 8.80 - 38.9.F_n$$

$$LCB = -13.5 + 19.4.C_P$$

Here LCB is estimated as percentage of length, positive forward of amidships.

7.1.4. *Tugboat propulsion*

Tugboat engines typically produce 500 to 2,500 kW ($\sim$ 680 to 3,400 hp), but larger boats (used in deep waters) can have power ratings up to 20,000 kW ($\sim$ 27,200 hp). Tugboats usually have an extreme power: tonnage-ratio; normal cargo and passenger ships have a PT-ratio (in kW: GRT) of 0.35 to 1.20, whereas large tugs typically are 2.20 to 4.50 and small harbor-tugs 4.0 to 9.5. The engines are often the same as those used in railroad locomotives, but typically drive the propeller mechanically instead of converting the engine output to power electric motors, as is common for diesel-electric locomotives. For safety, tugboats' engines often feature two of each critical part for redundancy.

Its engine's typically states a tugboat's power horsepower and its overall bollard pull. The largest commercial harbor tugboats in the 2000s–2010s, used for towing container ships or similar, had around 60–65 tons of bollard pull, which is described as 15 tons above "normal" tugboats.

Tugboats are highly maneuverable, and various propulsion systems have been developed to increase maneuverability and increase safety. The earliest tugs were fitted with paddle wheels, but these were soon replaced by propeller-driven tugs. Kort nozzles have been added to increase thrust per kW/hp. This was followed by the nozzle-rudder, which omitted the need for a conventional rudder. The cycloidal propeller was developed prior to World War II and was occasionally used in tugs because of its maneuverability. After World War II, it was also linked to safety due to the development of the Voith Water Tractor, a tugboat configuration which could not be pulled over by its tow. In the late 1950s, the Z-drive or (azimuth thruster) was developed. Although sometimes referred to as the Aqua master or Schottel system, many brands exist: Steerprop, Wärtsilä, Berg Propulsion, etc. These propulsion systems are used on tugboats designed for tasks such as ship docking and marine construction. Conventional propeller/rudder configurations are more efficient for port-to-port towing.

The Kort nozzle is a sturdy cylindrical structure around a special propeller having minimum clearance between the propeller blades and the inner wall of the Kort nozzle. The thrust-to-power ratio is enhanced because the water approaches the propeller in a linear configuration and exits the nozzle the same way. The Kort nozzle is named after its inventor, but many brands exist.

A recent Dutch innovation is the Carousel Tug, winner of the Maritime Innovation Award at the Dutch Maritime Innovation Awards Gala in 2006. The Carousel Tug adds a pair of interlocking rings to the body of the tug, the inner ring attached to the boat, with the outer ring attached to the towed ship by winch or towing hook. Since the towing point rotates freely, the tug is very difficult to capsize.

The Voith Schneider propeller (VSP), also known as a cycloidal drive is a specialized marine propulsion system. It is highly maneuverable, being able to change the direction of its thrust almost instantaneously. It is widely used on tugs and ferries.

From a circular plate, rotating around a vertical axis, a circular array of vertical blades (in the shape of hydrofoils) protrude out of the bottom of the ship. Each blade can rotate itself around a vertical axis. The internal gear changes the angle of attack of the blades in sync with the rotation of the plate, so that each blade can provide thrust in any direction, very similar to the collective pitch control and cyclic in a helicopter.

Chapter 8

Trawlers

8.1. Introduction

Although the term trawler is generally associated with fishing, it encompasses a few other interpretations as well. There are different types of trawlers that are used across the world. Many of these trawlers are not known widely among the general public.

The few main varieties of trawlers can be explained as follows:

8.1.1. *Fishing trawler*

As mentioned above, the fishing trawler is the most commonly known variant of the shipping vessel. This particular trawling system uses fishing nets that are attached to the trawler. The nets spread in the water mechanically and are propelled when the trawler moves.

In other words, the fishing net used by a fisherman is dragged in the water not by physical force but with the help of the mechanized trawler. Because of the usage of trawlers in today's times, it can be said that fishermen find fishing a more productive activity than what it used to be before.

8.1.2. *Recreational trawler/trawler boats/trawler yachts*

These trawler types are the lesser known variety of the naval vessel. Generally known as the recreational trawler or cruise trawler, trawler yachts are basically boats built in the form of trawlers. In some areas, such trawlers are also known as trawler boats. The main feature of these trawler varieties is the fact that they have a higher

Figure 8.1: Bottom Trawler [Phys.org, 2017]

engine capacity and are designed for the purpose of comfort and luxury. Their main duty is to provide recreation and relaxation. By using terminologies like trawler yachts and trawler boats, there is a distinguishing point provided so as to help laymen avoid using wrong trawler terms.

It has to be noted that since trawler yachts and trawler boats are mainly used as passenger naval vessels, the speed of the trawler increases considerably. Fishing trawlers have an average speed of about seven to nine knots, while the speed of the trawler yachts is nearly the double of that.

Apart from the aspect of speed, another major feature of the trawler boats that their design is subject to customization; which means that they can be designed and constructed the way a person wants them to be. Around the world, several such well-known trawler yacht builders create unique and excellent varieties of trawler boats and trawler yachts.

8.1.3. *Naval trawlers*

The concept of naval trawlers was very popular during the World Wars' time. Naval Trawlers were so named because they were designed in the shape of trawlers. The major feature and advantage of a naval trawler was the fact that they were bulkier and were capable of successfully launching underwater missiles and mines at the enemy vessel. This trawler variety is no longer famous owing to the wide-scale development in military engineering and armaments.

In addition to these to trawler types, there is also a concept known as bottom trawling. **Bottom trawling** is similar to the first variety of trawlers i.e. fishing trawlers.

In contemporary times, trawlers are more of a necessity than an item of comfort. They are exceedingly helpful to fisher folk as they help in catching fish faster and in large amounts. In addition, considering that there are trawler yachts and trawler boats that provide recreational feasibility to patrons, it is seen that the concept of a trawler can be used in many different ways. This multi-dimension of a simple concept is what makes trawlers very useful and alluring simultaneously.

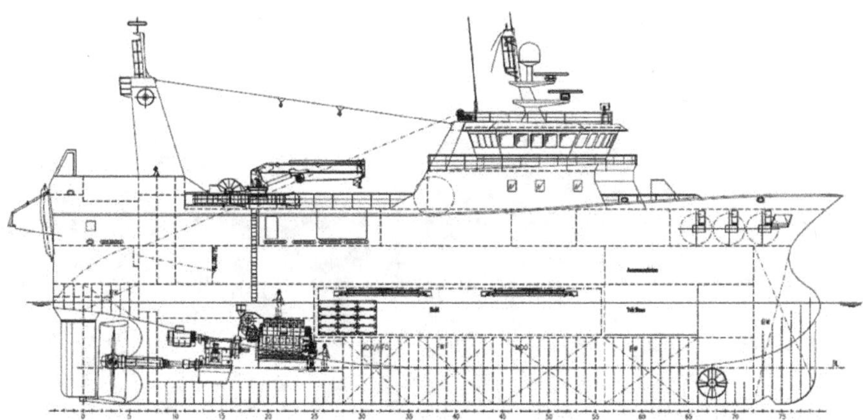

Figure 8.2: Image of a fishing trawler [Xiamen C&D Shipbuilding]

Table 8.1: Hydrostatics of a Trawler [Abdi Kukner and Mushin Aydm, "Influence of Design Parameters on Vertical Motions of Trawler Hull Forms in Head Seas"]

Parameter	Value
L	11–14 m
C_B	0.4890–0.5675
C_P	0.5900–0.6500
C_M	0.6625–0.7500
Velocity Range	7–9 knots (Fishing Trawler)
	14–18 knots (Trawler Yacht)
B/D	1.9–2.7
L/B	4.4–5.8

8.1.4. *Seakeeping criteria*

The performance of a fishing vessel particularly defines the success of its design. This is due to the fact that the ship has to survive in extreme weather, has to carry out her mission in all weather conditions by maintaining her speed and course with minimum discomfort to her crew.

Chapter 9

SWATH Ships

9.1. Introduction

A SWATH ship is the abbreviated form of the word 'Small Water plane Area Twin Hull' ship. Swath ships are designed in such a way that they have two or dual hulls instead of the conventional single hull. Technically, a swath ship is a type of a catamaran.

The two hulls of the swath ship are built so as to offer the maximum amount of balance or buoyancy to the ship. The placement of the hull is such that it rests under the water surface, making it safer and easier for the ship to sail when high tides and fast currents hit with full force. Whereas in single hull ships or boats, the hull floats above the surface of the water which makes the sailing difficult especially during unpredictable and rough seas.

The position of the hull under the surface of the water is also sometimes referred to being 'submarine submerged' as the technology looks similar to a submarine moving under the water. Along with the factor of buoyancy, a swath vessel is also known for its speed. Another feature and advantage of a swath vessel over the traditional ships is that it offers a bigger and wider area than the conventional types of ships.

One of the main reasons that act against the speed of a ship is the resistance of the ship's hull against the water. Reducing the contact of ship's hull with that of sea water also reduces the friction, allowing the ship to travel faster. SWATH technology, also known as the Small Waterplane Area Twin Hull, is one such method for increasing the speed of the vessel by reducing the craft's contact with vessel.

Figure 9.1: Carbon Catamaran [MTU]

9.2. Design and construction

Unlike the conventional hull, which rides on the water surface, the hull of the Swath vessel is fully submerged inside the water. However, that doesn't mean that the area of hull in contact with the water is more. The peculiar shape of SWATH is such that its hull, which has a very less area, remains completely submerged in water, whereas the deck rides high above the water surface, thus providing higher speed with excellent stability to the craft.

The water plane of a ship is that part that actually cuts the water as the ship moves forward. The trick for minimizing friction is to reduce the surface area of the water plane by keeping it as small as possible. This allows the vessel to have high stability even in the roughest weather condition.

Swath vessel consists of two submarine shaped hulls attached to its bottom area. The shape of a Swath vessel resembles that of a catamaran. The two submarine shaped hulls are attached to the vessel using a thin structure or struts. The main aim of the entire design is to keep the volume of that part of the boat which is inside the sea water and has the highest wave energy acting, the least.

9.2.1. *Additional design aspects*

It is to note that during the construction process, if the contact area of the strut with the water is decreased, the transverse area between the two hulls should be increased in order to acquire appropriate transverse stability to resist the heeling movement. Also, smaller struts will provide less space for machinery inside the craft and also make the lower hulls less accessible. Thus, it is very important to choose the right water plane area for the overall performance of Swath Vessel.

Also, in order to smoothly ride over the small waves in coastal areas, the clearance underside of the Swath vessel should be adequate. This particular movement of Swath, which involves small vertical motions, is called platforming.

Swath ships are used as both — cruise as well as cargo ships and are also used as army and research vessels. Fredrick G. Creed, a Canadian, first created the swath ship in the year 1938. The patent for the same was obtained in the year 1946 in Great Britain. The vessel, however, was utilized for the first time in the shipping industry only in the 60s and 70s. The first swath ship was used as a research vehicle than as a transporting vessel across the water.

Some of the swath ships that are in operation today include the Cloud X, which acts as a ferry between the Bahamas and Florida and the USNS Impeccable — a US marine surveillance vessel.

One of the main disadvantages of swath ships is that the two hull build of the ship makes it far more power and energy consuming than the single hull variety ships. Because of this, another disadvantage of the ship also appears regarding the overall expenditure involved in the construction and energy supply of the swath ships. The energy required to be spent on a swath vessel is eight times or 80% more than the energy spent on a ship with a single hull.

Swath technology is an efficient way of cruising at higher speed than those of the conventional crafts without sacrificing the necessary

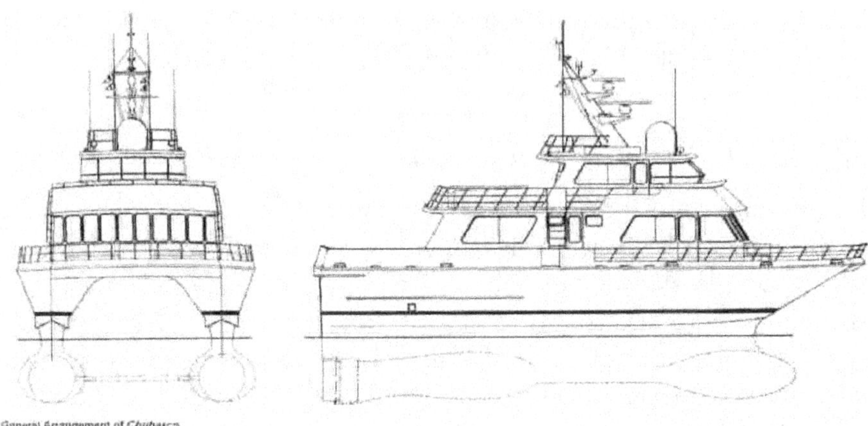

Figure 9.2: 'Chubasco' — built in 1986 by Ocean Systems Research Ltd — marine grade aluminium [Solar Navigator]

stability. In their own way, swath ships have revolutionized water transport. In spite of their faults, they have been a unique presence and will continue to do so in the days to come. Figure 9.1 shows a SWATH ship.

9.3. Important parameters

Table 9.1: Summary of Parameter Range of Trawlers (American Bureau of Shipping)

Parameter	Range
Length	20–100 m
L/B	1.75–4.75
L/H	2.25–6.75

Chapter 10

Surface Effect Ships

10.1. Introduction

Surface effect ships are special types of vessels, which are a combination of hovercrafts and catamarans. They are so designed that they can use both air cushion and conventional propulsion as and when required. This allows them to have greater speed on seawater.

Incorporating the technology of both catamarans and hovercrafts, surface effect ships have been a part of the maritime domain since the 60s. Initially designed to be used as naval vessels, these types of ships, slowly over the years gained extreme favorability as vessels to be used for non-naval purposes too.

The main advantage of the surface effect ships is that they offer stability in any kind of water conditions because of being equipped with both — dual hulls and air springs. Need arising, the double hulls of the vessel are complemented by the air springs, which support and cushion the vessel providing it with a good balance.

These types of ships have hulls on their sides, which calls for optimum stability while being utilized in the high seas. Originally, these vessels were designed to be operational only in smaller water parts, but in contemporary times, they have started to be utilized in the bigger water areas as ferry vessels.

Alongside stability and efficient functionality, speed is yet another factor that adds to the singularity of these vessels. When the first surface effect vessels were being designed, the target speed intended for these vessels was placed at 80 knots. Also considering that these

                    Principles of Marine Vessel Design

Figure 10.1:   Surface Effect Ship [Marine Insight]

vessels were intended to be naval vessels, speed became a very vital component.

During the late 20th century, developed nations like the United States, Norway, the United Kingdom and the Soviet Union, revolutionized the concept of surface effect vessels, as each country came up with distinctive surface effect vessels. In the present times, the Norwegian naval patrols make extensive utilization of these types of ships.

A couple of United States' surface effect ships were also used to launch missiles between the 1960s and 70s. The vessels, named as SES100A and SES100B were built as a part of a practical experiment conducted by the American military experts.

- Both vessels measured 24 meters lengthwise, with a breadth of 12 meters and a tonnage of 100 tones.
- Their speed offered by the vessels almost touched 100 knots, almost exceeding the naval target of 80 knots.
- At the time of the missile launching, the speed attained was around 60 knots.

Popular names associated with the surface effect vessel engineering are Hover marine of the United Kingdom and the Rohr Industries of the United States.

Further researching has been taking place in the domain of surface effect vessels so as to better the technology and to make the vessels operational in any given water area. In the days to come, it can be expected that the vessel's area of operation will be further widened, as will be its operational scope.

Table 10.1 gives some parameters of a 2200 ton SES designed for the US Government.

Table 10.1: Important Parameters of a 2200 tons SES [Surface Effect Ships, US Department of Navy]

Parameter	Value
Gross weight	2200 tons
Payload	250 tons
Overall length	62.484 m
Overall beam	30.004 m
Cruise speed	80 knots
Cruise range	4000 nautical miles in sea state 3
Propulsion engines	25000 hp
Lift Engines	12500 hp maximum

Chapter 11

Submarines

11.1. Introduction

Submarines are underwater self-propelled crafts that are designed and built to perform underwater operations for a stipulated amount of time. Submarine design consists of a single or double hull system that houses all the necessary systems and manpower required for completion of their mission. This, though, is a very simple description of a very complex engineering product, which are used for a wide range of purposes such as underwater research, underwater rescue, and submarine warfare; the last one being the most widely used.

The primary submarine design objectives are:

- The submarine should cater to the functional purpose of the customer.
- The design should be capable of being constructed with the available resources.
- The cost of the project should be acceptable by the customer.

11.2. Parts of a submarine

11.2.1. *Outer hull and pressure hull*

Most designs of submarines have two hulls. The light hull ("casing" in British usage) of a submarine is the outer non-watertight hull which provides a hydrodynamically efficient shape. The pressure hull is the inner hull of a submarine that maintains structural integrity with the difference between outside and inside pressure at depth. The hull that houses all the accommodation spaces, weapons, weapon control

Figure 11.1: USS Indiana on Sea Trials [Marine Insight]

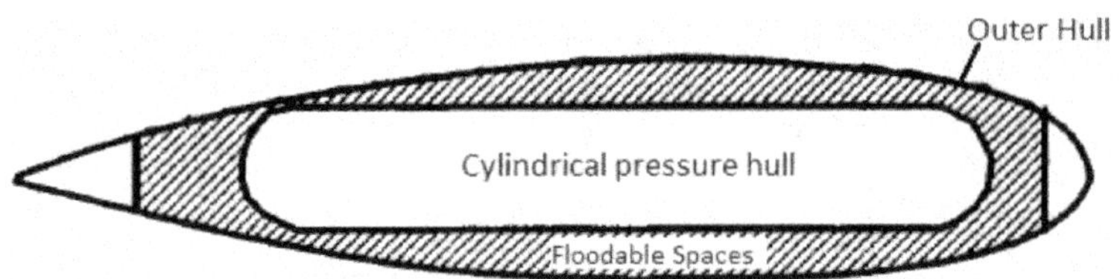

Figure 11.2: Cylindrical Pressure hull and Outer hull of a submarine [Marine Insight]

systems, communications and control room, battery banks, main and auxiliary machinery, is the pressure hull. It is called the pressure hull because it is designed to withstand the hydrostatic pressure at the maximum operable depth of the submarine.

The pressure hull is housed inside the outer hull, which is not pressure tight. In submerged condition, the spaces between the outer and the inner hull always remain flooded with seawater. Hence, the hydrostatic pressure on the outer hull is negligible.

11.2.2. *Main Ballast Tanks (MBTs)*

The "floodable" spaces are compartmentalized into tanks, which in submarine terminology, are called Main Ballast Tanks. The distribution of main ballast tanks in a submarine depends on the shape and interaction of the outer and pressure hull. Ballast tanks, with the diving planes, give a sub control over the submarine's buoyancy, particularly during the first part of a dive or a return to the surface from the depths. When the ballast tanks are filled with air, the submarine rises to the surface because it has positive buoyancy. With water inside the tanks, the sub has negative buoyancy so it sinks deeper into the ocean. The tanks at the front (known as the front trim tanks) are usually filled with water or air first, so the submarine's front (bow) falls or rises before its rear (stern). The ballast tanks can also be used to help a submarine surface very quickly in an emergency. Some designs have MBTs only at the forward and aft regions, and the rest of the pressure hull is flushed with the outer hull. Other designs have completely different outer and pressure hull, with space for ballast between them. Some arrangements of MBTs are shown in the figures below.

Figure 11.3: Exposed pressure Hull (MBTs at forward and aft) [Marine Insight]

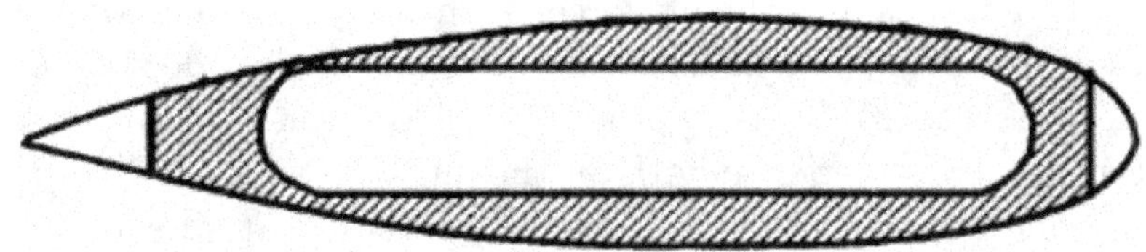

Figure 11.4: Enclosed Cylindrical Pressure Hull (MBTs throughout the length) [Marine Insight]

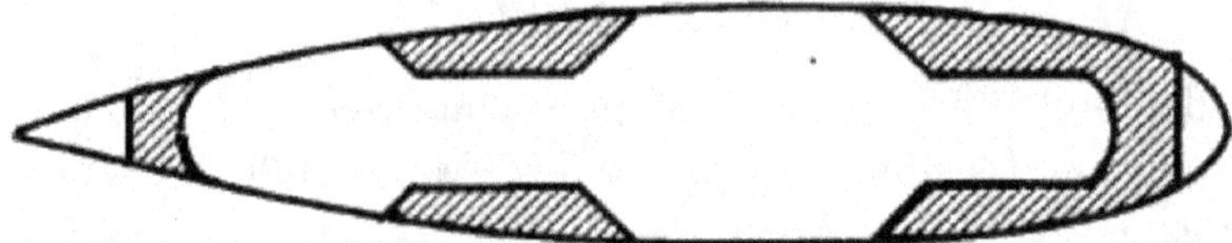

Figure 11.5: Waisted Pressure Hull (MBTs at certain parts of the length)[Marine Insight]

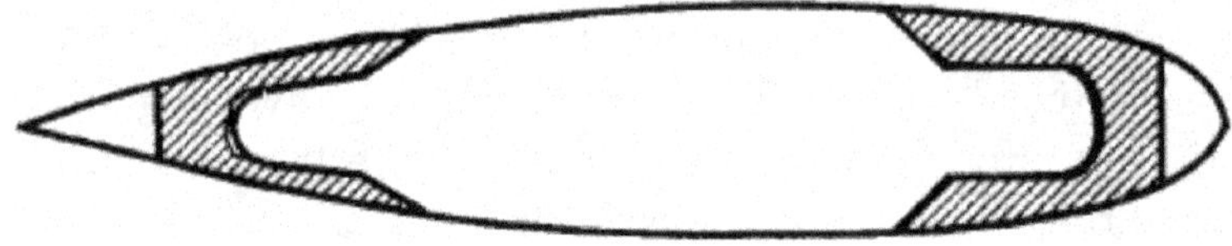

Figure 11.6: Exposed pressure hull reduced at ends (MBTs at forward an aft) [Marine Insight]

11.2.3. *Control surfaces/planes*

Submarines have fins called diving planes or hydroplanes. When the submarine is in submerged condition, changes in direction and depth is carried out by use of hydroplanes that act as control surfaces. They work a bit like the wings and control surfaces (swiveling flaps) on an airplane, creating an upward force called lift. Buoyancy is the tendency of something to sink, rise, or float at a certain depth. While it's underwater, a submarine is negatively buoyant, which means it tends to sink, left to its own devices, if it's not moving. But as the submarine's propellers push it forward, water rushes over the planes, creating an upward force called lift that helps it remain at a certain depth, creating a state of neutral buoyancy (floating). The planes can be tilted to change the lift force, so making the submarine climb or dive through the sea, as necessary. The planes provide most of the submarine's control of its depth, most of the time. The amount of lift they generate depends both on the angle to which they're tilted and on the submarine's speed (just as the lift that wings generate depends on a plane's speed and "angle of attack"). Unlike a surface ship, submarines are subjected to lesser heave and pitch motions due to absence of surface wave effects. A pair of hydroplanes or fins at the forward and aft are used to control the heave and pitch independently. Two hydroplanes mounted at the aft in the vertical plane are used to change the lateral direction of the submarine when in motion. These are basically called rudders. It is important to note that unlike ships, the rudders of a submarine are forward of the propeller. In case of a ship, the rudder requires the propeller outflow for maximum lift efficiency. But in a submarine, since the entire hull is submerged, undisturbed streamlined flow is incident onto the rudder surface. If the submarine rudder were placed aft of the propeller, the flow onto the rudder would be more turbulent, increasing the probability of cavitation. Also hydroplanes operate at optimum efficiency only at high speeds.

11.2.4. *Sail or bridge fin*

The sail (American usage) or fin (European/Commonwealth usage) of a submarine is the tower-like structure found on the dorsal (topside) surface of submarines. Submarine sails once housed the conning tower (command and communications data center), the periscope(s), radar and communications masts (antenna), though most of these functions have now been relocated to the hull proper (and so the sail is no longer considered a "conning tower"). In layman's term it's a tall tower packed with navigation and other equipment at the centre of a submarine. The profile of the bridge fin in a submarine design is always an aerofoil shape; it acts as a hydrofoil as the submarine sails with just the fin above water. This shape reduces the drag on the submarine.

11.3. Engine

Gasoline engines and diesel engines used by cars and trucks, and jet engines used by planes, need a supply of oxygen from the air to make them work. Things are different for submarines, which operate underwater where there is no air. Most submarines except nuclear ones have diesel-electric engines. The diesel engine operates normally when the sub is near the surface but it doesn't drive the sub's propellers directly. Instead, it powers an electricity generator that charges up huge batteries. These drive an electric motor that, in turn, powers the propellers. Once the diesel engine has fully charged the batteries, the sub can switch off its engine and go underwater, where it relies entirely on battery power.

Early military submarines used breathing tubes called snorkels to feed air to their engines from the air above the sea, but that meant they had to operate very near the surface where they were vulnerable to attack from airplanes. Most large military submarines are now nuclear-powered. Like nuclear power plants, they have small nuclear reactors and, since they need no air to operate, they can generate power to drive the electric motors and propellers whether they are on the surface or deep underwater.

11.4. General arrangement of a submarine

Figure 11.7 shows the arrangement of different hulls and other components of a dolphin submarine.

From the above figure we can understand the basic arrangement of a submarine. The pressure hull and outer hull are clearly distinguishable in the above figure of the submarine design. The forward part of the pressure hull houses the weapon systems and sensors. The sensors are usually housed in the flooded space between the forward of the pressure hull and the outer hull. Sensors are always placed at

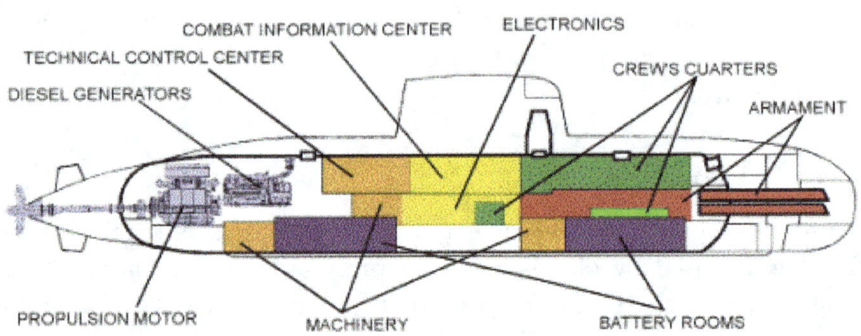

Figure 11.7: General Arrangement of a Dolphin Submarine [Dolphin Project, Israel Tripod]

the forward for reduction of noise from turbulent flow at the aft and obstruction of machinery in case of aftward position.

The midship section of the pressure hull is mainly used for the following purposes:

- Ship and Weapon Control Systems
- Accommodation and Life Support
- Battery bank
- Machinery and Auxiliary Machinery
- Propulsion Compartment

11.5. Influence of some general hull parameters

Systematic model tests with streamlined bodies of revolution having fineness ratios similar to those of submarines were performed at the David Taylor Model Basin (now Naval Surface Warfare Center) in Washington, Gertler (1950), Landweber and Gertler (1950). The offsets of the models were derived from a 6th degree polynomial. The investigation varied the following parameters:

Fineness ratio L/D
Prismatic coefficient $CP = \nabla/(\pi.0.25 \ D2.L)$
Nose radius (nondimensional) $r_0 = R_0.L/D2$

The results of the investigations together with statements taken from Arentzen and Mandel (1960) are summarised in the tables given below. The data in the tables is taken from Volker Bertram's "Submarine Hull Design".

11.5.1. *Fineness ratio*

The optimum L/D depends to a small extent also on the Reynolds number Rn, Granville (1976). According to model tests on the DTMB series 58, the optimum would be for hulls of the type 1400 (approximate bare hull volume $1800 \, \text{m}^3$) at L/D $\approx$ 7 for max. speed and L/D $\approx$ 6.5 for slow speed. For practical reasons (e.g. space requirements), a fineness ratio of L/D = 7 is scarcely met. Fineness ratios

Table 11.1: Influence of L/D ["Submarine Design", Volker Bertram]

Value of L/D	Result
L/D > 9	Residual Resistance is constant
L/D = 6.5 (= 7 if control surfaces are taken into account)	Optimum value for bare hull

of L/D = 9 can be considered as low already for submarines. As the total resistance curve over L/D is rather flat, there is little penalty involved in moving to such fineness ratios.

11.5.2. *Prismatic coefficient*

The prismatic coefficient is defined for submarines as

$$C_P = \nabla/(\pi/4.D^2.L)$$

where ∇ is the volume of the envelope and D the maximum hull diameter. The prismatic coefficient has a significant influence on the resistance.

Gertler (1950) found $C_P = 0.61$ as optimum for bodies of revolution having constant volume and constant L/D. Compared with $C_P = 0.70$ the resistance is less by 9%, compared with $C_P = 0.55$ by 13%. Considering also control surfaces does not materially alter the position of the optimum C_P.

Special care is required in the detailed shaping of the lines for greater C_P values. The following table lists favourable combinations of C_P, $C_{P,E}$ and L'_X.

C_P is the total prismatic coefficient, $C_{P,E}$ is the prismatic coefficient of the residual hull without the parallel midbody, L'_X the length of the parallel midbody, L'_X/L, where L is the overall length.

11.5.3. *Nose radius*

Typical value of nose radius for a submarine is 2.46.

Table 11.2: Favourable combinations of C_P, $C_{P,E}$ and L'_X [Arentzen and Mandel (1960)]

C_P	$C_{P,E}$	L'_X
0.60	0.600	0.000
0.64	0.612	0.068
0.68	0.625	0.143
0.70	0.632	0.185
0.72	0.638	0.225
0.76	0.652	0.311
0.80	0.667	0.400
0.84	0.682	0.495
0.88	0.700	0.600

11.5.4. *Speed*

The speed of the submarine is very important and it can be achieved by a given power output developed by the propulsion plant. Most submarines and torpedoes can't go much faster than 40 knots ($\sim$ 46 mph). Higher speeds are possible, but it requires so much power that it's not really feasible (torpedoes only have so much fuel).

Chapter 12

Catamaran

12.1. Introduction

A catamaran is a multi-hulled watercraft featuring two parallel hulls of equal size. It is a geometry-stabilized craft, deriving its stability from its wide beam, rather than from a ballasted keel as with a monohull sailboat. A traditional sailboat is a monohull — in other words, it has only one hull centered on a heavy keel. A catamaran is balanced on two hulls, with the sails in the middle. Being ballast-free and therefore lighter than a monohull, catamarans often have a shallower draft (draught) than a comparably-sized monohull. Depending on the size of the boat, the space separating the two hulls might be filled by a cockpit, a main cabin, and usually some netting. The two hulls combined also often have a smaller hydrodynamic resistance than comparable monohull, requiring less propulsive power from either sails or motors. The catamaran's wider stance on the water can reduce both heeling and wave-induced motion, as compared with a monohull, and can give reduced wakes. Catamarans range in size from small (sailing or rowing vessels) to large (naval ships and car ferries). The structure connecting a catamaran's two hulls ranges from a simple frame strung with webbing to support the crew to a bridging superstructure incorporating extensive cabin and/or cargo space.

12.2. Performance characteristics

Catamarans have two distinct primary performance characteristics that distinguish them from displacement monohull vessels: lower

Figure 12.1:   General Catamaran [The Multihull Company]

Table 12.1:   Catamaran Parameters [Molland *et al.* 1994]

Parameters	Range
L/B	6–12
B/T	1–3
C_B	0.33–0.45
S/L	0.2, 0.3, 0.4, 0.5
F_n	0.2–1.0

resistance to passage through the water and greater stability (initial resistance to capsize). Choosing between a monohull and catamaran configuration includes considerations of carrying capacity, speed, and efficiency.

Table 12.1 shows some important parameters of a catamaran.

12.2.1. *Stability*

Catamarans rely primarily on form stability to resist heeling and capsize. Comparison of heeling stability of a rectangular-cross section monohull of beam, B, compared with two catamaran hulls of width

Table 12.2: Hydrostatics of catamaran of Equal Displacement Model [Srikanth Asapana, "Resistance Prediction for Asymmetrical Configurations of High Speed Catamaran Hull forms]

Displacement [Kg]	16.4
Wetted Surface Area [m^2]	0.597
Draft (main)	0.107
Block Coefficient	0.383
L/B Ratio	4.74
B/T Ratio	2.719
$L/\nabla^{1/3}$	5.729

Table 12.3: Hydrostatics of Asymmetrical Catamaran Model [Srikanth Asapana, "Resistance Prediction for Asymmetrical Configurations of High Speed Catamaran Hull forms"]

Displacement [Kg]	16.4
Wetted Surface Area [m^2]	0.597
Draft (main)	0.107
Block Coefficient	0.383
L/B Ratio	8.33
B/T Ratio	1.794
$L/\nabla^{1/3}$	6.264

$B/2$, separated by a distance, $2 \times B$, determines that the catamaran has an initial resistance to heeling that is seven times that of the monohull. Compared with a monohull, a cruising catamaran sailboat has a high initial resistance to heeling and capsize — a fifty-footer requires four times the force to initiate a capsize than an equivalent monohull.

12.2.2. *Resistance*

At low to moderate speeds, a lightweight catamaran hull experiences resistance to passage through water that is approximately proportional to the square of its speed. A displacement monohull, by comparison, experiences resistance that is at least the cube of its speed. This means that a catamaran would require four times the power in order to double its speed, whereas a monohull would require eight times the power to double its speed, starting at a slow speed. For powered catamarans, this implies smaller power plants (although two are typically required). For sailing catamarans, low forward resistance allows the sails to derive power from attached flow, their most efficient mode — analogous to a wing — leading to the use of wing-sails in racing craft.

12.2.3. *Trade offs*

One measure of the trade-off between speed and carrying capacity is the displacement Froude number (Fn_V), compared with *calm water transportation efficiency*. Fn_V applies when the waterline length is too speed-dependent to be meaningful — as with a planing hull. It uses a reference length, the cubic root of the volumetric displacement of the hull, V, where u is the relative flow velocity between the sea and ship, and g is acceleration due to gravity:

$$Fn_V = u/(\sqrt{g}.V^{1/3}).$$

Calm water transportation efficiency of a vessel is proportional to the full-load displacement and the maximum calm-water speed, divided by the corresponding power required. Large merchant vessels have a Fn_V between one and zero, whereas higher-performance powered catamarans may approach 2.5 — denoting a *higher speed* per unit volume for *catamarans*. Each type of vessel has corresponding calm water transportation efficiency, with large transport ships being in the range of 100–1,000, compared with 11–18 for transport catamarans — denoting a *higher efficiency* per unit of payload for *monohulls*.

Chapter 13

Trimaran

13.1. Introduction

A trimaran is a multihull boat that comprises a main hull and two smaller outrigger hulls (or "floats") which are attached to the main hull with lateral beams. Most trimarans are sailing yachts designed for recreation or racing; others are ferries or warships. Trimarans are particularly known in relation to sailing craft. They perform a number of roles ranging from frigates to container ships. Research into the suitability of trimaran forms for the large vessel applications seems to have begun only in earnest towards the end of the 20th century with results beginning to appear regularly in the open literature in the 1990's.

These trimaran designs developed for large vessels are essentially a variant on the slender monohull concept. Figure 13.2 gives a typical example with the design of a coastal passenger ferry. With the addition of two smaller outrigger side hulls for stability the centre hull can be made even more slender in the endeavor to reduce the wave making resistance in particular even further. As a trimaran will have a greater wetted surface area than a monohull of equivalent displacement then at low speeds where viscous drag dominates, it will have a higher total resistance.

Figure 13.1: General Trimaran [The Boat Works]

Table 13.1: Model Principal Characteristics [Ackers *et al.* (1997)]

	Main Hull	Each Side Hull
LOA	1.77 m (5.81 feet)	0.643 m (25.32 in)
LWL	1.689 m (5.543 feet)	0.6096 m (24 in)
Beam — max	(4.94 in)	(1.0 in)
Beam — WL	0.1254 m (4.74 in)	0.0254 m (1.0 in)
L/B	14	24
Draft	0.06228 m (2.425 in)	0.0254 m (1.0 in)
Wetted surface	0.2435 m^2 (2.622 ft^2)	0.03218 m^2 (0.3464 ft^2)
Displacement	11.92 lb	0.37 lb

	Trimaran
Wetted surface	0.30795 m^2 (3.3148 ft^2)
Displacement	12.66 lb

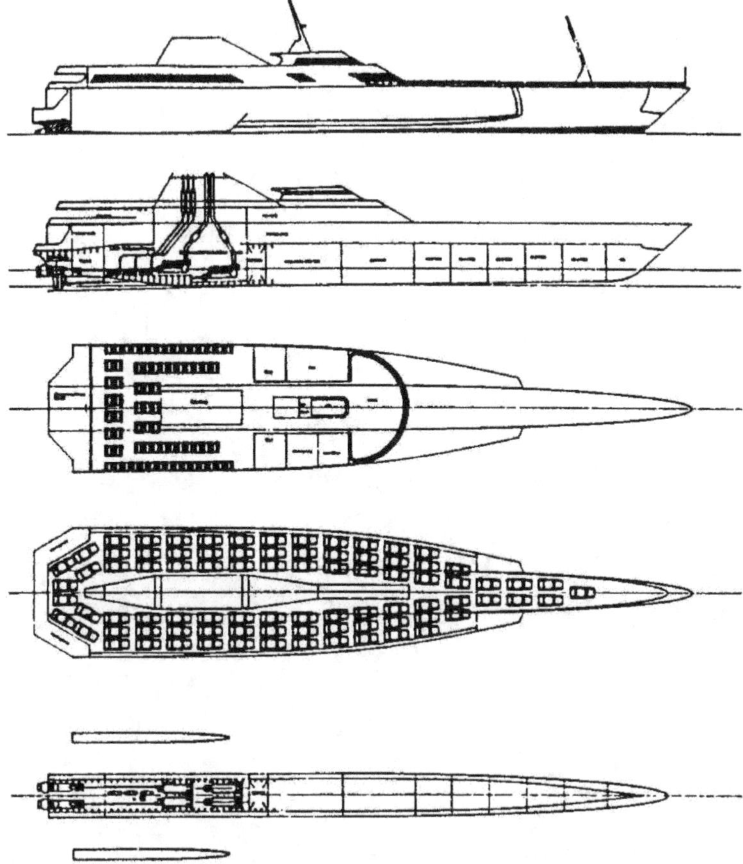

Fig. 7 General Arrangement of Canadian Coastal Ferry

Figure 13.2: Canadian Trimaran Design [Pattison & Zhang (1995)]

Chapter 14

Pentamaran

14.1. Introduction

A **pentamaran** is a multihull vessel with five hulls, a further extension of the ideas behind trimarans (three hulls) and catamarans (two hulls). As with the most promising trimaran concepts, this concept is also based on a long slender main centre hull. In this case this main hull is instead stabilised by the addition of four smaller pontoons. The concept was developed to meet the requirements of a 13000 tonne vessel, able to travel at 30 knots with an installed power in the order of 30 MW. Comparable existing designs had required 50 MW.

As mentioned in "Preliminary Review of Novel Hull forms", preliminary study discounted the use of catamarans and symmetric trimarans due to excessive drag. A slender monohull gave the lowest drag but was unstable. An asymmetric trimaran (i.e. with side hulls smaller than the centre hull), proved the practicable optimum. In this case stability requirements were achieved with only a slightly higher drag resulting than for the monohull. The use of lift devices such as foils and air cushions was assessed but found unsuitable as they resulted in an overall increase in powering required due to the speed range of interest (Froude number of the main hull of 0.32). Bulbous bow and stern appendages were also investigated with positive results.

14.2. Types of pentamaran

There are two primary types of pentamaran.

Figure 14.1: Applications of a Pentamaran Concept [WRK]

14.2.1. *Parallel type*

The parallel pentamaran consists of five almost equally narrow hulls in a side-by-side fashion, as in the M80 Stiletto, also known as an M-Hull. Historically, ships have evolved to become narrower and deeper to achieve speed and stability. The M Hull however becomes wider, because its distinctively wide hull captures the vessel's bow wave and redirects the energy under the hull. These pentamarans can be very fast when planning yet quite stable in rough seas for a small ship. In addition, these ships offer plenty of space due to their width, although they do not have a lot of space below decks. Their shallow draught also makes them suited for shallow waters.

14.2.2. *Outrigger type*

Outrigger type pentamarans may be likened to an evolved trimaran, with a long and narrow central hull at least 50 metres long, though none with this configuration have ever been built, and two pairs of outriggers — one pair aft, one pair amidships — as opposed to the trimaran's single pair. Going with the current trend of fast conventional hulls in naval shipbuilding, the pentamaran is based upon a

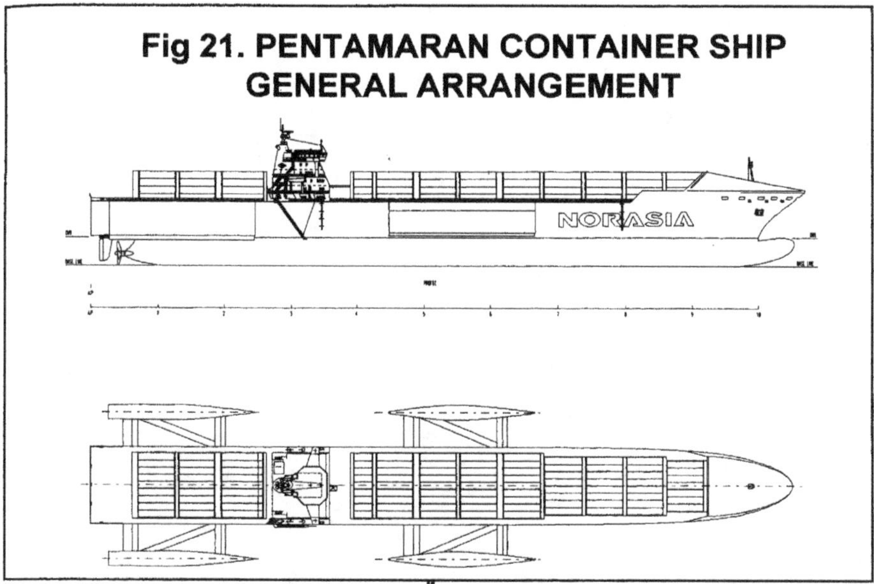

Figure 14.2: The Pentamaran Concept [Gee *et al.* (1997)]

narrow but long hull, that is deeper than it is wide, so that it cuts through waves. The outriggers then provide the stability such a narrow hull needs. While the aft sponsons act as trimaran sponsons do, the front sponsons do not touch the water normally; only if the ship edges to one side they do, in which case the sponsons will slow the ship and level it up until the ship is stable again. BMT Group, a shipbuilding and engineering company located in the United Kingdom, has proposals for a fast cargo ship and a yacht using this kind of hull.

Chapter 15

Air Cushioned Vehicles (ACVs)

15.1. Introduction

Air Cushioned Vehicles are collectively known as Ground Effect Machines (GEMs). Most of the GEMs are ACVs of the peripheral jet type. In general usage by the industry, the popular press, and most of the technical press, the word hovercraft has come to be synonymous with ACV.

Air-cushion vehicle (ACV) craft is designed to travel close to but above ground or water. It is also called **Hovercraft.** These vehicles are supported in various ways. Some of them have a specially designed wing that will lift them just off the surface over which they travel when they have reached a sufficient horizontal speed (the ground effect). Others are supported by fans that force air down under the vehicle to create lift. In a plenum chamber vehicle the rate of leakage of this air from underneath the vehicle is reduced by placing a skirt around the lower edge of the craft. In an annular jet vehicle the rate of leakage is reduced by directing the air downward and inward from the outer edges of the vehicle. Air propellers, water propellers, or water jets usually provide forward propulsion.

Air-cushion vehicles can attain higher speeds than either ships or most land vehicles and use much less power than helicopters of the same weight. Air-cushion suspension has also been applied to other forms of transportation, in particular trains, such as the French Aerotrain and the British Hover train. A relatively smooth land or water surface, however, is a necessity; most of these vehicles cannot clear waves higher than 1–1.67 m.

Figure 15.1: The Hovercraft [Access Science (2017)]

The first recorded design for an air-cushion vehicle was put forward by Emmanual Swedenborg, a Swedish designer and philosopher, in 1716. The project was short-lived and a craft was never built, for Swedenborg soon realized that to operate such a machine required a source of energy far greater than any available at that time. In the mid-1870s, the British engineer Sir John Thornycroft built a number of model craft to check the air-cushion effects and even filed patents involving air-lubricated hulls, although the technology required to implement the concept did not yet exist. From this time both American and European engineers continued work on the problems of designing a practical craft.

In the early 1950s the British inventor Christopher Cockerell began to experiment with such vehicles, and in 1955 he obtained a patent for a vehicle that was "neither an airplane, nor a boat, nor a wheeled land craft." He had a boat builder produce a two-foot prototype, which he demonstrated to the military in 1956 without arousing interest. Cockerell persevered, and in 1959 a commercially built one-person Hovercraft crossed the English Channel. In 1962 a British vehicle became the first to go into active service on a 31-km ferry run.

15.2. Specifications

The maximum size of air-cushion vehicles is now over 100 tons; some of them travel at over 160 km per hr. Although air-cushion vehicles of several thousand tons have been under development for many years, it is in small vehicles, usually called flarecraft that the greatest current potential market exists; current flarecraft can carry one to eight people at 240 km per hr.

Peter Du Cane (Author of 'High Speed Small Craft') suggests that when dealing with marine hovercraft, it is more useful to describe the craft by their amphibious or non-amphibious nature than their air cushion systems. There are three types of marine hovercrafts namely, amphibious, non-amphibious and hybrid hovercraft.

15.2.1. *Amphibious hovercraft*

The typical amphibious hovercraft possesses a full peripheral jetted (or similar) skirt, air screw propulsion using gas turbines and lightweight construction derived from aircraft and flying boat practice. Typical modern operational craft of this type include the BHC series of craft.

15.2.2. *Non-amphibious hovercraft*

The non-amphibious craft is characteristically a sidewall vessel powered by marine diesel or gas turbines driving water propulsors and employing a comparatively cheaper though still lightweight form of construction. The HM2 and American SES craft are important examples of this type of craft.

15.2.3. *Hybrid hovercraft*

The Vosper Thornycroft (VT) represents the 'hybrid' or 'semi amphibious' hovercraft. It is a full peripheral skirted hovercraft with aircraft type gas turbine engines driving super cavitating water screws. The craft is constructed from marine grade aluminium alloys, but the structural design is based on lightweight aircraft engineering practice.

For a given engine power the amphibious hovercraft is the fastest type in calm water conditions and in such conditions probably represents the fastest present type of marine craft that can be constructed. However as well as being the fastest it is also the most expensive type of hovercraft.

15.3. Hovercraft design considerations

For the preliminary design of a new hovercraft one takes into account the data accumulated from previous successful designs and the information collected from the technical and environmental surveys.

There are two major aspects which must be considered when one starts to develop any new hovercraft design; namely, the operational environment and the craft characteristics. These aspects assume particular relevance in different areas of the preliminary design work.

When a hovercraft service is being proposed for a particular route one of the first things to be considered is the craft seakeeping ability. This is very much influenced by the cushion depth, which is often related to the 'significant wave height'. Generally the following relation can be used for finding the maximum wave height.

$$H_{max} = 1.6 H_{1/3}$$

where H_{max}: Maximum wave height

$H_{1/3}$: Significant wave height

A typical commercial requirement is for cushion depth maximum waves not to be exceeded for more than 10 percent of the operational time. The cushion depth required to satisfy this condition can be obtained from a "Percentage Exceedance Diagram" for the particular area under consideration.

The craft payload and speed may be embodied in a requirement submitted by a potential customer, or they may be laid down by the hovercraft departmental management in the form of a feasibility study. In either event, it is necessary to determine whether these parameters are reasonable when related to the general craft concept. In the case of a passenger/car ferry craft for example, a useful design index is that of "work capacity", defined as the product of the payload in tons and service speed in knots.

Knowing the payload to be carried, a first order estimation of the all up weight (AUW) may be found from the percentage payload characteristics of craft similar in design to that projected. Most of the hovercrafts have percentage payloads in the range of 20 per cent to 30 per cent of the AUW. The possession of a very high percentage payload is necessarily essential, as this can only be achieved by employing a very light weight and hence expensive structural design. The percentage payload must be considered in terms of the potential economic performance of the craft.

The choice of cushion pressure depends mainly on the payload density, previous craft experience and upon the skirt system to be employed. There is a general increase in the cushion pressure with increase in craft size. The types of payloads which a commercial hovercraft may carry are passengers, cars and freight. In military fields it may be required to carry a large amount of fuel for long range; the percentage disposable load is then important.

There are a number of requirements that must be kept in mind when designing a passenger or car craft, the most important of which are the regulatory safety requirements. When a passenger/car craft is being designed, consideration must be given to the passenger/car ratio.

Noise is another craft characteristic which is of great importance in commercial and military fields. For commercial craft a low external noise level is important because craft tends to operate between centres of population. For military craft external noise may be important, whether it is transmitted through the air or through water, depending on the craft role.

The dominant source of noise is the propulsion device and this must be selected to suit the role. Water propelled craft have the lowest airborne noise level. Air propelled craft have no problems with underwater noise and the open air propeller gives the worst airborne noise level.

15.4. Power requirements

The following calculations and results are taken from Peter Du Cane's "High Speed Small Craft". The intention is to give the reader the order of magnitude and a sense of proportion between the various values of the drag components, without going into detail or lengthy mathematical expressions.

15.4.1. *Lift power*

There are several theories for calculating the lift power of a peripheral jet cushion. The consideration which decides the amount of cushion air flow required is derived from wave pumping considerations. Waves passing beneath the craft cause a rearward displacement of air within the cushion. To prevent a "plough-in" under this sweeping or pumping action the lift system must be designed to maintain the design cushion characteristics. Therefore, to provide a satisfactory ride characteristic and a reasonable lift horsepower the volume flow is calculated from the following equation.

$$C_{Q,S} = Q/S.\sqrt{\rho/2}.P_C$$

where $C_{Q,S}$: non dimensional coefficient

S: cushion volume flow

P_C: cushion pressure

P: air density

This value is usually determined from experience of optimum performance of lift power and skirt drag. The current range of values is

0.010 to 0.020, but to large extent is dependent on cushion depth. The order of the value of $C_{Q,S}$ can be obtained from

$$C_{Q,S} = h/5^L.$$

where h: cushion depth

L: cushion length

It should be remembered that this is a suggested minimum requirement for $C_{Q,S}$ and wherever possible higher values would give better ride characteristics. Also, a higher $C_{Q,S}$ tends to reduce skirt drag.

The lift horsepower per ton of all up weight (AUW) is given by

$$LHP/ton = (118.(H_f/P_C).C_{Q,S}.P_C^{1/2})/\eta_f,$$

where H_f: fan total head

η_f: fan total efficiency (approximately 0.8–0.85)

Although dependent on the shape and type of fan intake, for all practical purposes the lift power can be assumed to be the same whether the craft is statically hovering or at forward speed.

15.4.2. *Propulsive power*

For a hovercraft to move forward at a speed V, it requires sufficient power or thrust to overcome the total drag, and the component of this drag on a hybrid type of hovercraft are:

$$D_M = \text{Momentum drag}$$

$$D_O = \text{Aerodynamic Profile drag}$$

$$D_W = \text{Induce wave drag}$$

$$D_K = \text{Skeg drag}$$

$$D_{ns} = \text{Net skirt drag}$$

$$D_{wav} = \text{Drag due to waves}$$

Momentum Drag (D_M)

A continuous air suppying is required to maintain the cushion; the momentum drag arises from the change in momentum of the free stream air, which relative to the craft is brought to rest as it enters the fan inlet. If a hovercraft is proceeding at speed the required free stream air relative to the craft speed has to be brought to rest. This momentum drag can be written as

$$D_M = \rho Q V,$$

where Q: volume flow

$$V: \text{craft velocity.}$$

Aerodynamic Drag (D_O)

All moving bodies are subjected to a profile drag which is dependent on the shape and surface finish of the body. A streamlined shape will tend to have a low profile drag and a bluff shape a high drag. The profile drag is usually expressed in terms, of the non-dimensional coefficient $C_{D,O}$ where

$$D_O = C_{D,O^{1/2}}.\rho.V^2.A_f,$$

where A_f: frontal area.

$C_{D,O}$ has typically values in the range from 0.25 to 0.5.

Induced Wave Drag (D_W)

Hovercraft moving over water generate a pattern of induced waves which radiate away from the craft and this results in a drag force on the craft.

Skeg Drag (D_K)

The total skeg drag is usually obtained by the summation of various drag components of profile, wave and spray drag. However in its simpler form the skeg drag is given by

$$D_K = C_K.0.5.\rho W.V^2.S_K,$$

where C_K: nondimensional coefficient

$$\rho_W: \text{water density}$$

$$S_K: \text{wetted area of skeg and rudder}/2.$$

Net Skirt Drag (D_{ns})

This is the drag associated with the contact between the skirt design, cushion pressure, craft size and speed. Because of the difficulty of a theoretical assessment, this term is obtained by model tests.

Rough Weather Drag (D_{wav})

The effect of wind on the drag of a hovercraft influences those drag components dependent on airspeed; namely momentum and aerodynamic drags, and these can be readily calculated.

The effect of waves on the drag of a hovercraft is to increase the skeg and skirt drags. The former can be roughly calculated by adding to the skeg profile area and extra area which is considered to be time averaged, taking into account the wave length and height.

Total Drag (D_T)

It is the summation of all the component drags. The thrust curve is then plotted on the drag curve and where the two intersect, this gives the speed of the craft. The thrust curve is obtained from

$$T = (HP.550.\eta)/V,$$

where

$$T: \text{Thrust}$$

$$HP: \text{Horsepower at the propeller}$$

$$H: \text{Propulsive efficiency.}$$

For a super cavitating CPP the propulsive efficiency is in the region of a 55–60 per cent at the design cruise speed, dropping to 25–30 per cent at the shallow water hump speed.

Chapter 16

Conclusion

We live in time of continuous technological development, and every new ship that is built should in some way contribute to this ongoing technical advancement. Yet in practice it is not so easy to design a vessel that is fast, economical, environmentally friendly, and also user-friendly. Shipyards want a ship that is light and easy to build. Ship owners want a vessel that has a record-breaking total capacity, is easy to maintain and operate, is inexpensive and safe, and so on. Shippers and cargo owners want the cargo to safely reach its destination. Masters and crew are looking for a ship that is easy to operate, comfortable, and also safe, in the broadest sense of the word, and especially with respect to the working environment. So, the design of vessels must be such that it meets these expectations in the best possible way.

Appendix

Evaluating Wetted Surface Area:

In the initial design stage when lines plan is unavailable but basic parameters of the vessel are known then several different formulations can be applied to evaluate the wetted surface area which have been illustrated in equations 11 to 29. All areas are in m^2 unless stated.

Mumford	$S = 1.7Ld + \dfrac{\nabla}{d}$		(11)
	$S = (1.7T + C_B B)L_{WL}$	Without appendages	(12)
Taylor	$S = C\sqrt{\nabla L}$	$C \cong 2.58$	(13)
Froude	$S = \Delta^{2/3}\left[36.38 + \dfrac{1.636L}{\Delta^{\frac{1}{3}}}\right]$	L in ft and Δ in tons	(14)
Unknown	$S = L[1.52d + (0.374 + 0.85C_B^2) \times B]$	In ft^2	(15)
Unknown	$S = \dfrac{\nabla}{B}\left[\dfrac{1.7}{C_B - 0.2(C_B - 0.65)} + \dfrac{B}{d}\right]$		(16)
Lap (NSMB)	$S = (3.4\nabla^{1/3} + 0.5L_{WL})\nabla^{1/3}$	Without appendages	(17)
Afonasiew	$S = 1.02(0.8C_B + 0.2)(B + 2T)L_{WL}$	With rudder + keel	(18)
Danckwardt	$S = \dfrac{\nabla}{B}\left(\dfrac{1.7}{C_B - 0.2(C_B - 0.65)} + \dfrac{B}{T}\right)$	General cargo and passenger vessels	(19)

	$S = \dfrac{\nabla}{B}\left(\dfrac{1.7}{C_B} + \dfrac{B}{T}\left(0.92 + \dfrac{0.92}{C_B}\right)\right)$	Without appendages for trawlers	(20)
Riehn	$S = 0.9L_{WL}(2T + C_W B)$	Coastal ships without appendages	(21)
Sabit	$S = \nabla^{2/3}\left(3.432 + 0.305\dfrac{L_{BP}}{B} + 0.443\dfrac{B}{T} - 0.643C_B\right)$	Without appendages	(22)
Mercier-Savitsky (1973)	$\dfrac{S}{\nabla^{2/3}} = 2.262\left(\dfrac{L}{\nabla^{1/3}}\right)^{0.5}\left(1 + 0.046\left(\dfrac{B}{T}\right) + 0.00287\left(\dfrac{B}{T}\right)^2\right)$		(23)
Holtrop	$S = L(2T + B)\sqrt{C_M}\left[0.453 + 0.4425C_B - 0.2862C_M - 0.003467\dfrac{B}{T} + 0.3696C_{WP}\right] + 2.38\dfrac{A_{BT}}{C_B}$	A_{BT} is the transverse section area of bulb	(24)
Schneekluth and Bertram (1998) [17]	$S = (3.4\nabla^{1/3} + 0.5L_{WL})\nabla^{1/3}$		(25)

AMECRC Series [18]

The wetted surface area is valid for L/B (4 to 8), B/T (2.5 to 4) and
C_B (0.4 to 0.5) as per Multiple Regression Analysis and Non-linear
Estimation of AMECRC series of high-speed semi-displacement
round bilge hull forms are shown in equations (26) and (27) respec-
tively:

$$\frac{S}{\nabla^{2/3}} = 3.328344 + 0.74494 \left(\frac{L}{B}\right)^{2/3} (C_B)^{-2/3}$$

$$+ 0.35227 \left(\frac{B}{T}\right) (C_B)^{-2/3}$$

$$+ 0.04630664 \left(\frac{L}{B}\right)^{2/3} \left(\frac{B}{T}\right) C_B^{-1}$$

$$- 0.0379448 \left(\frac{L}{B}\right)^{4/3} C_B^{-1}$$

$$- 1.367162 \left(\frac{B}{T}\right)^{1/3} (C_B)^{-1/3} \tag{26}$$

$$\frac{S}{\nabla^{2/3}} = 0.889635 + 0.0667 \left(\frac{L}{B}\right)^{0.35274} \left(\frac{B}{T}\right)^{1.47721} C_B^{-0.90566}$$

$$+ 1.64397 \, C_B^{-0.4423} \left(\frac{L}{B}\right)^{0.3812} \tag{27}$$

SKLAD Series (L/B 4–8, B/T 3–5, C_B 0.35–0.55) [18]

$$C_S = \frac{S}{\nabla^{2/3}}$$

$$C_S = 2.456288 + 0.434391 \left(\frac{L}{B}\right)^{-2/3} \left(\frac{B}{T}\right)^{2/3} (C_B)^{2/3}$$

$$- 0.013612 \left(\frac{L}{B}\right)^{-4/3} \left(\frac{B}{T}\right)^{-1} (C_B)$$

$$+ 0.000188 \left(\frac{L}{B}\right)^{-3} \tag{28}$$

NPL and S-NPL Series [18]

$$C_S = 4.44579 + 0.25272 \left(\frac{L}{B}\right)^{2/3} \left(\frac{B}{T}\right)$$

$$- 18684148 \left(\frac{L}{B}\right)^{-2} \left(\frac{B}{T}\right)$$

$$- 14.46315 \left(\frac{B}{T}\right)^{-1} + 131.2228 \left(\frac{L}{B}\right)^{-2} \left(\frac{B}{T}\right)^{2/3}$$

$$+ 23.01607 \left(\frac{B}{T}\right)^{2/3} - 0.00813 \left(\frac{L}{B}\right)^{2} \left(\frac{B}{T}\right)$$

$$+ 0.528452 \left(\frac{L}{B}\right)^{2/3} \left(\frac{B}{T}\right)^{2/3} \tag{29}$$

Sahoo Prasanta, Preliminary Ship Design

Sahoo Prasanta, Resistance Prediction Methods for High Speed Crafts

References

Abdi Kukner and Muhsin Aydm, "Influence of Design Parameters on Vertical Motions of Trawler Hull Forms in Head Seas".

AccessScience. (2017). Retrieved from https://accessscience.com

Anish. (2016). Image General Bulk Carrier. Retrieved April 2, 2019, from https://www.marineinsight.com

Brock Vergakis,. (2018). "5 Facts About the Navy's Newest Submarine USS Indiana". Retrieved from https://pilotonline.com

Diesel Duck. (2011). The Birth of the Bulker. American Bureau of Shipping's Surveyor. Retrieved from http://www.dieselduck.info

Doros A Argyriadis, "Modern Tug Design with Particular Emphasis on Propeller Design, Maneuverability, and Endurance B".

Foxwell, D. (2017). Image Vietnamese gig for Solstad Construction support vessel. AME Maritime. Retrieved from https://www.amemaritime.com

Genco Shipping & Trading. (2015). Retrieved from https://www.gencoshipping.com/

Gesamtverband der Deutschen Versicherungswirtschaft. (2009). Stowage Factor. Retrieved from http://www.tis-gdv.de

Hannah Hickey, "Bottom-trawling techniques leave different traces on the seabed". Retrieved form https://phys.org/news

Haun, E. (2016, November). Image Anchor Handling Tug Supply Vessel. MarineLink. Retrieved from https://www.marinelink.com

Hub, F. (2016). Image MSC Zoe. Retrieved from https://freighthub.com

Image Bourbon dive support vessel. (2017). Retrieved from http://www.bourbonoffshore.com

Image Cell Guides. (2014). Retrieved from http://www.containerhandbuch.de

Image Daleel Platform Supply Vessel. (2015). Retrieved from https://www.scmdaleel.com

Image Operation of self-unloading bulk carriers. (2010). Retrieved August 2, 2019, from http://bulkcarrierguide.com

Image Ramform Viking. (2019). Retrieved from https://www.pgs.com

Ingpen, B. (2015). Types of Bulk Carriers. Retrieved from http://maritimesa.org

International Maritime Organization. (2014). Design Process & Constraints.

Juhari Husin and Zainal Ashririn Shahardin, "A Simplified Method of Calculating Propeller Parameters for Small Trawlers" (1984).

Kairis, S. (2012). Ships Geometry and Hull Definition. Retrieved from https://officerofthewatch.com

LNG tanker and specialized vessels' gain could be tankers' loss. (2013). SeaNews Turkey. Retrieved from https://www.seanews.com.tr

Marex, "Med Marine Delivers Tugboat to Turkey's Izmit Bay". Maritime-Executive. Retrieved from https://www.maritime-executive.com

Marine Insight. (2019). Retrieved from https://www.marineinsight.com/

Marine Notes. (2013). Retrieved from http://marinenotes.blogspot.com

Marine Wiki. (2019). Retrieved from https://en.wikipedia.org/wiki/Marine

Martopo and Soegiyanto. (2009). Ship Unloading Equipment. Retrieved from http://greatseafarertraining.blogspot.com

Jan P. Michalski, "Statistical data of hull main parameters useful for preliminary design of SWATH ships", Gdansk University of Technology, ul. Narutowicza 11/12, 80-952 Gdańsk Polish Naval University, ul. Śmidowicza 69, 81-103 Gdynia.

Offshore Supply Vessels. (2019). Retrieved from https://www.exxonmobil.com

A. Papanikolaou, "Selection of Main Dimensions and Calculation of Basic Ship Design Values", Ship Design, DOI 10.1007/978-94-017-8751-2_2, Springer.

Dr Prasanta Sahoo, "Preliminary Ship Design".

Dr Prasanta Sahoo, "Preliminary Review of Novel Hull Forms".

Peter Du Cane. (2007). "High Speed Small Craft"

PRIF, Unit 7, "Container Ships".

Sharda, "What are Surface Effect Ships", Marine Insight, Retrieved from https://www.marineinsight.com

US Department of Navy, "Surface Effect Ships".

Volker Bertram, "Submarine Design".

Wärtsilä, "Encyclopedia of Marine", TecP. (2017).

William F. Clifford, and James P. Cummings. (1993); "Design and Operational Experience of the SWATH Ship "Navatek I Ludwig H. Seidl", Marine Technology, Vol. 30, No. 3, pp. 153–171.

Index